Space Station
FRIENDSHIP

Space Station FRIENDSHIP

A Visit with the Crew in 2007

Dick Lattimer

STACKPOLE BOOKS
Lanham • Boulder • New York • London

Published by Stackpole Books
An imprint of The Rowman & Littlefield Publishing Group, Inc.
4501 Forbes Boulevard, Suite 200, Lanham, Maryland 20706
www.rowman.com

Unit A, Whitacre Mews, 26-34 Stannary Street, London SE11 4AB

Distributed by NATIONAL BOOK NETWORK

Copyright © 1988 by Stackpole Books
First Stackpole Books paperback edition 2017

British Library Cataloguing in Publication Information Available

Library of Congress Cataloging-in-Publication Data

The Hardback edition of this book was previously cataloged by the Library of
Congress as follows:

Lattimer, Dick.
     Space Station Friendship.

     Bibliography: p.
     Includes index.
     I. Title.
PS3562A775S64   1988   813'.54   87-18138
ISBN 0-8117-1683-X

ISBN 0-8117-1683-X (cloth)
ISBN 978-0-8117-3699-2 (paper : alk. paper)
ISBN 978-0-8117-6666-1 (electronic)

Printed in the United States of America

For Mom and Dad

And for my wife and best friend, Alice,
and all the other teachers in the world
who sacrifice each day
in a world of chalk dust and runny noses
for the coming millennium

Ellison Onizuka

LTC Ellison S. Onizuka, mission specialist
Space Shuttle Orbiter mission 51-Lima,
the last flight of *Challenger*

Ellison Onizuka was a friend of mine. Our common bond was his interest in my sport of archery and my interest in his work in the space program. El was an amateur archer and loved to hunt and fish.

The last time we talked was just a couple of weeks before El died aboard the space shuttle *Challenger*. He called to tell me how much he and his family were enjoying my book *All We Did Was Fly to the Moon*. On his first flight he and his crewmates had taken several archery items into orbit for us for our Fred Bear Museum, and I suggested that someday he might also take up a copy of my book for display in the Museum. Whether he had a copy with him that frigid January morning I have never really tried to find out. The idea has always been too painful for me. El's laughter and friendship have never been far from my mind as I researched and wrote this book.

As we go to press there are still many decisions up in the air in Washington and elsewhere about our space station. I would like to share with you a thought that El once expressed: "I dare you to dream . . . then go after that dream with sweat and commitment."

Contents

Foreword_____

Now is the time to take longer strides, time for a great new American enterprise, time for this Nation to take a clearly leading role in space achievement, which in many ways may hold the key to our future on Earth.

—President John F. Kennedy

The space station was still far in the future when I kicked the lunar dust off my boots and climbed back into *Challenger* that December day in 1972. Yet it seems like just yesterday that President Kennedy's words excited all of us in the astronaut corps and people around the world. At the time, the goal seemed almost impossible, really beyond our reach. Yet during the Mercury, Gemini, and Apollo programs that followed, we struggled and failed and tried again until finally we figured out how to do the impossible.

When Jack Schmitt and I lifted off the lunar surface in the ascent stage of *Challenger,* none of us in the aerospace business could ever have guessed that it would be so very long before we would return. After all, we had broken a trail, and Americans have always been curious to see what lies at the end of new trails. But perhaps there really is a tidal pace to man's achievements. Maybe we do need to stop and take our breath as a species. Luckily the space shuttle was already on the drawing boards, and that sustained our space vision for a while. Once more we were laughing and working in weightlessness. But as in our moon program, people soon lost interest in the shuttle adventures, and we became complacent.

Our friends that were lost on board space shuttle *Challenger* woke us up to the realities and the true costs of human destiny, just as the crew of *Apollo 1* did many years before. The changes brought about by *Apollo 1* ultimately led to successful lunar landings and discoveries. The legacy of *Challenger* was much the same.

The space station provides us with a focus for our human achievement. As with Apollo and the shuttle, it gives us what Norman Cousins once called a great gift. He asked what it might be worth to us to take a peek at civilization 1,000 years or more in the future. "The great gift possessed by humans," he said, "is the ability to accelerate their development in a given period of time. . . . The space program offers an opportunity to avoid going through a tunnel of 1,000 years before coming to a point we all regard as representing a higher station in life. . . . I am deeply interested in improving the human condition. I see the space program as a means of doing just that."

Our space station gives us what amounts to a celestial campus in which to make educated guesses and trial-and-error mistakes. When Dick Lattimer and I were in college back in the 1950s, he at Indiana University and I up the road at rival Purdue, neither of us could foretell that there would ever even be a space station. In those days NASA itself had not even been formed. Yet today both of us take for granted that mankind will reach as far out into the cosmos as we care to go.

No longer is it a question of "Can we do it?" Now it seems more a matter of "When will we want to do it?" When will we once more experience that surge of mankind's universal urge to "press on," as we say in the astronaut office. The space station will give us a platform from which to launch those coming expeditions into infinity. Yet at the same time it will provide us with a test bed for the improvement of our daily lives here on Earth.

I remember so vividly that last day I spent on the moon. In a photograph Jack took of me holding the American flag in the Valley of Taurus-Littrow, over my left shoulder is a barely recognizable blue-and-white ball, no larger than a baseball. It was our planet Earth, and home seemed so very far away.

I said something then to the effect that with all the wonderful things we were achieving at that moment, "there might be Someone else that has something to do with it, too, and I've been reading His signs, maybe not from Him directly, but His in spirit. And if He's listening, I'd like to thank Him, too. Godspeed the crew of *Apollo 17.*"

Today, all of us on Earth need to renew our commitment to search for a better life. It's as simple as that. Forget all the engineering terms, the biotechnology lingo, and the chemistry formulas. Yes, we need people trained in all of them for our voyages aboard the space station. But we need to conquer our restlessness. And we need to get on with it.

Houston, Texas   Gene Cernan
                                                 Captain, USN (Retired)

Gene Cernan, commander of Apollo 17, was the last man to leave his footprints on the moon. It would be 20 years before America would take its next major step in space by building its orbiting island space station. As Cernan left the surface of the moon in 1972 he spoke of the future. He asserted that "America's challenge of today has forged man's destiny of tomorrow." Surely that destiny includes hope for mankind as the benefits of living and working together in space are realized.

Acknowledgments_____

Special thanks must go to all the scientists, researchers, and writers who provided the original research material upon which most of this book was based. Some of their contributions are listed in Further Reading. Wherever possible I have given their names in the text so that you can look further into their research at your leisure.

Next, I must thank the people within the National Aeronautics and Space Administration across America who helped so very much in providing me with background material, reference photography, and artwork. They are listed here by location: *NASA Headquarters, Washington, D.C.:* Richard E. Halpern, director, utilization division, office of space station; Dr. B. J. Bluth, NASA clinical psychologist, program system engineering and integration division; Dr. Bevan French, scientist, solar system exploration division; Bill O'Donnell and Mark Hess, office of space station; and Lee Saegesser, NASA history office. *Johnson Space Center, Houston, Texas:* Astronaut Joe Henry Engle, commander of the space shuttles *Columbia, Discovery,* and *Enterprise;* Sylvia Stottlemeyer, astronaut office; and Terry White, Lisa Vazquez, and Mike Gentry, public affairs office. *Marshall Space Flight Center, Huntsville, Alabama:* Luther Powell, office of space station; Terry Eddleman, public information office; and Charles Lindbergh

Yeager, Micro-Craft, Inc. *Ames Research Center, Moffett Field, California:* Peter Waller, public affairs office; and Marc M. Cohen, space human factors office. *Kennedy Space Center, Florida:* Carl DeLaune, engineering and integration office; Nancy Gunter, astronaut office; and Dr. Wick Hoffler, medical office. *Jet Propulsion Laboratory, Pasadena, California:* Jurrie Van Der Woude, public information office. *Goddard Space Flight Center, Greenbelt, Maryland:* Janet K. Wolfe, public affairs director.

People from the international aerospace community were also very helpful. They include: *Canada:* John Wildgust, National Research Council, Ottawa. *Europe:* Ian Pryke, European Space Agency, Washington, D.C. office. *Japan:* Chu Ishida, National Space Development Agency of Japan, Washington, D.C. office. *United Kingdom:* John E. Humby, British Aerospace PLC, Stevenage, Herts.

Thanks also to those from the American aerospace community: Bill Rice, Boeing Aerospace Company, Seattle, Washington; Dick Barton and Sue Cometa, Rockwell International, Downey, California; Susan Flowers, McDonnell Douglas Astronautics Company, St. Louis, Missouri; Jeff Fister, McDonnell Douglas, Huntington Beach, California; William J. Ennis, TRW Electronics and Defense Sector, Redondo Beach, California; Bill Woodruff, Optoelectronics Division, Honeywell, Inc., Richardson, Texas; Evan McCollum, Martin Marietta Michoud Aerospace, New Orleans, Louisiana; Glen Talcott, Martin Marietta, Huntsville, Alabama; Jan Wrather, Lockheed Missiles & Space Company, Sunnyvale, California; Sidney Lerman, Bendix Aerospace, Teterboro, New Jersey; and Montye C. Male, TRW Electronics & Defense Sector, Redondo Beach, California.

Several other people were also a great help to me during my research. They include: Sue Stephenson, Inertial Confinement Fusion, Lawrence Livermore National Laboratory, University of California, Livermore; Freda Carr (mother of *Skylab* commander Jerry Carr), Balboa Island, California; Dr. Ralph Grams, Medical Systems Division, University of Florida School of Medicine, Gainesville; Dr. Alvin Moreland, School of Veterinary Medicine, University of Florida, Gainesville; Don Eyles, The Charles Stark Draper Laboratory, Inc., Cambridge, Massachusetts; Helen Weeks, Kealakekua, Hawaii; Charles Michael Haller, Horizon Productions, Ltd.; Dick Mauch, Bassett, Nebraska; Fred Bear, Gainesville, Florida; Robert F. Kelly, Gainesville, Florida; and the staff of the University of Florida Library.

Special thanks to Peggy Senko, my editor at Stackpole, for the initial invitation to step across the threshold into tomorrow and for her pleasant encouragement along the way.

And finally, special thanks for their patience and encouragement must go to two typists who helped me during the various stages of this book, Nancy Walker and Jo Meskimen, Gainesville, Florida.

Introduction _____

Gardeners of the Universe

As I write these words, I am reminded that this week is the anniversary of the loss of the *Challenger* and her crew. Writing this book has been therapy for me since they left Earth. It has made me pause and turn my head from that white boiling "Y" cloud of the past toward the exciting planets that await us in our future.

I am reminded of this past-future business in another way today. In my hand I hold the ankle bone of an ancient long-jawed elephant. I spotted this strange-looking rock one spring morning as I chased wild turkeys with a bow and arrow along Plum Crick. Four and a half million years ago the bone's owner grazed in the rich vegetation of subtropical Nebraska. One day she lay down, shuddered, and was still. A million years later a young couple walked across an African plain at Laetoli. The girl paused to look back over her shoulder as a volcanic eruption rained warm ash down upon them. Their footprints were frozen there in time.

Today our *Pioneer 10* spacecraft is out there somewhere searching for a tenth planet in our solar system. It has long since passed beyond the orbits of all nine known planets and by 1991 we shall have an answer. In 10,507 years *Pioneer 10* will pass Barnard's Star, 5 billion miles away. In the year 26,135 Proxima Centauri will be off its port bow. After 852,075 years, still traveling at a speed of 30,000 miles per hour, it will come within 4.1 light years of D + 25 1495, a solar-type star. All of these stars could have planets like our own.

With *Pioneer*, we are leaving our intellectual spore throughout the universe. Many scientists expect it to outlive our own solar system. Most of us do not know these things just as the long-jawed elephant and the young African couple did not know that they, too, were leaving spore to mark their passing.

I believe that the orbiting space station that the United States is about to construct holds the promise of sending us on our physical journey through the stars. I call the station *Friendship* after our first orbital U.S. spaceship, John Glenn's *Friendship 7*. To me the name reflects what should be the ultimate goal of all of our undertakings in space.

I like to think of the space station as a life raft or an island in nearby outer space. In reality it will be a cluster of Thermos-bottle-like living and working modules attached to a gridlike framework that will support experiments, antennas, cameras, and solar arrays. A couple of free-flying platforms with scientific instruments attached will fly nearby, and a tugboat-type maneuvering vehicle will be used to service them and to do other work around the station.

Actually, our first permanent facility in space will be our second space station. Our first was *Skylab*, an orbital collection of leftovers from the Apollo moon program that was a home for three crews of astronauts in 1973 and 1974.

NASA calls the space station "an outpost in orbit around the Earth for the conduct of science and commercial enterprises and for the development of new technologies." The space station will be a campus, factory, planetarium, research hospital, laboratory, think tank, and cathedral in the sky, visible every moment of every day somewhere on Earth as the brilliant morning star of the next millenium.

The scientific data in this book is based on actual research. Space station *Friendship* reflects technology existing today and projected plans for the future. Who knows what new developments await us between now and the time the first space station elements rocket into orbit?

In this book we will visit *Friendship* in 2007, the fiftieth anniversary of *Sputnik*'s orbit of Earth. To help you project yourself on board the space station and better understand the work that will be done there, I've invited some fictional crew members to come along for the voyage. We'll join three rookies as they board the station for the first time. With them we'll visit each of the modules designed for our space station and observe the activity there during the first week of their stay.

The real names of many of the people who have contributed to the development of the space program are included in the *Friendship* story. The three rookies, the crews of *Friendship* and the other space vehicles, the Cornell professor who inspired Billy to study astronomy, and the space officials around the world who participate in the conference call in chapter 13 are all fictional.

Mary Two Hawks, Billy Wong, Wayne Morrison, and the other people who live on *Friendship* in these pages will someday exist. Their names and their backgrounds may be different, but people like them will one day add their smudged fingerprints to its walls.

According to NASA's plan, in 1989 the contractors chosen by NASA will start fabrication of the space station modules. In 1990 NASA will begin crew training, and in 1991 the contractors will deliver the first elements to NASA. In early 1994 we'll begin taking these pieces up to orbit on our fleet of space shuttles. And a year later the space station will reach a man-tended capability. That means we will be able to make short visits up to it. By July 4, 1995, a crew of at least eight people will be living there. By 1998 the station should be completed.

After that the space shuttle will visit the station at least every 45 days, or eight times a year. Crew rotation will occur every 90 days with four new crew members going up with each space shuttle flight. The annual operating cost of the space station will be $1.4 billion, and the station will have a 30-year life.

Johnson Space Center in Houston, Texas, will be in charge of the manned aspects of the station as well as the power, data management, and life support systems. Marshall Space Flight Center in Huntsville, Alabama, will oversee user operations in the on-board laboratories of the United States, the European Space Agency, and Japan. Kennedy Space Center in Florida will be responsible for logistic operations, and Goddard Space Flight Center in Maryland will be in charge of unmanned platforms.

I urge the White House and Congress to immediately approve the necessary funding for the development and deployment of a rescue lifeboat for the space station. The fear and anguish of our friends on board *Challenger* at 73 seconds after launch are more than enough justification for this investment in tomorrow's safety.

Thomas G. Pownall of Martin Marietta Corporation has said, "It is time to restart the engine. We have been in a parking orbit too long. While the USSR's space budget has been rising 15 percent a year, our current space budget is only one-fourth what it was back in the 1960s. The U.S. space program costs each American about a dime a day. We could double it without reaching the cost of a postage stamp. For an estimated two cents a day, we can build and maintain the space station."

It's fitting that the crew of our space station will be international since our nation is made up of people born in 155 different countries. America could have chosen to keep its space station all to itself. It says a great deal about us as a people that we choose to share this grand adventure with the other free peoples of the world.

A recent "Nova" television special revealed that many scientists believe they can trace us all back to the same African mother through our DNA. And a recent study at Florida Atlantic University suggests that all of us on Earth are no more distantly related than fiftieth cousins, and that most of us are a great deal closer than that. We are family. These two facts alone illustrate the folly of the way we live on this small planet.

Many people have suggested that reaching out into space may be a good way to channel the creative energies of those of us on Earth in a positive way rather than making war. I feel very strongly that the space station can provide us with just such a focus. Let us be the gardeners of the universe and seed vegetables, flowers, wildlife, and trees, along with our grandchildren, throughout the stars.

1

A Rendezvous
with Infinity_____

"Roger, *Friendship*. We copy. Five bye."

"Uh, yeah, *Discovery*. We confirm radar acquisition. One thousand eighty nautical miles."

"Do you have comm and tracking lock?"

"Roger, *Discovery*."

"Ladies and gentlemen, please return to your launch positions for the remainder of rendezvous and berthing. We need to keep our real estate on the flight deck clear, don't ya know."

Space shuttle *Discovery*, a veteran of more than 20 years of space voyages and now outfitted with her new payload bay doors, continued on her silent chase to the space station now being tracked by radar. The three passengers and four regular crew members floated back to their positions and strapped themselves in.

They saw nothing of the target ahead. The commander, the pilot, and the mission specialist on the flight deck were the only ones who could watch the slow, steady closing with *Friendship*, the giant space station sailing smoothly ahead.

On the deck below, the shuttle's passengers could hear only the cooling pumps and fans. And their own heartbeats, faster now with the excitement.

"*Discovery,* we confirm that guidance has you within our 30-degree cone on the minus velocity vector and you are go at this time."

"Roger, *Friendship.* Here comes the cold beer."

"Sounds good to me, Smokey. Bring 'er on in."

Time began to slow down for the passengers as their objective drew closer. They knew that they would never be the same again once they had boarded *Friendship.*

"*Discovery,* the checklist compels me to confirm you acknowledge that we have override capability to wave off rendezvous if station safety warrants it."

"Confirm, *Friendship.*"

"*Discovery,* we mark you now in Zone 4 for stable orbit rendezvous. Do you roger that?"

"You got it, *Friendship;* 100 nautical miles and closing."

There was no conversation now from behind the flight deck or from the deck below. For all the levity going on, this was serious business. Two objects speeding along at 18,000 miles per hour were going to meet at the same moment and place in space, at a point in infinity never before visited by man, there to couple and become one.

"We have you on our dish, Commander. Confirm good radar data."

"Roger, *Friendship.*"

"Twenty nautical miles. Laser docking switch on."

"Go for co-orbit zone requested."

"Go, *Discovery.*"

Now it was very quiet. Now *Discovery* would seek its offset point 1,000 feet ahead of *Friendship,* in its final proximity operations procedure.

"Confirm Prox Ops."

"Roger, *Discovery.*"

Discovery was now approaching *Friendship* from ahead with its payload bay doors open, exposing its radiators to shed excess heat. To those on the ground it would appear that *Discovery* was flying tail down, but of course there is no up or down in space. The flight deck crew could see the silent station through the windows above their heads.

Down on the mid-deck, where the three nervous space rookies now floated in their launch seats, Chico, a veteran mission specialist, broke the tension of the moment. "Well, boys and girls, what did you think of that liftoff?"

"Fantastic!"

"I'm still afraid to take a breath."

Chico laughed. "Yeah, you never quite get used to it. It's a shame

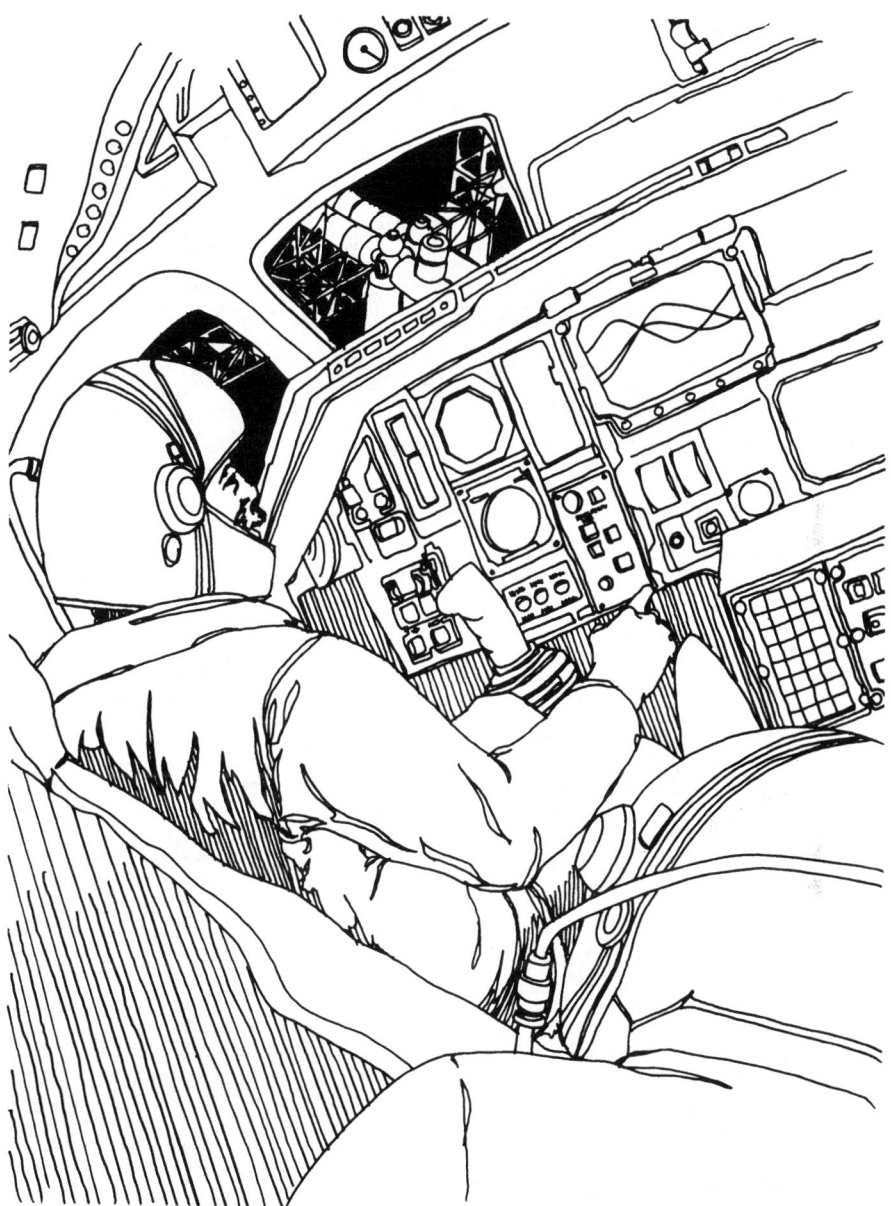

Approaching space station *Friendship* from the flight deck in the space shuttle *Discovery*.
To rendezvous, the shuttle must approach slowly, under and ahead of the orbiting space
station. It must be careful to miss the co-orbiting free-flying platforms and the tethered
experiments speeding along with *Friendship* in her orbit.

you're stuck down here where you can't see what all is going on, but there's just so much room upstairs in the attic, so we're relegated to the basement of this here machine."

Chico continued, filling in the rookies on what was happening on the flight deck.

The launch and rendezvous had all been planned by computer. *Friendship* was allowed to get just about overhead while *Discovery* sat on the pad. Once *Discovery* was up at orbital speed, *Friendship* was two or three miles ahead and about 1,000 feet above as planned. A burn put *Discovery* in circular orbit. Through small burns the velocity was slowed gently to bring *Discovery* up to the same altitude as *Friendship*.

"By now Smokey and Bob have moved from the front of the flight deck back to their rendezvous station at the back by the payload bay," Chico was saying. "They'll do all of the final approach from there by looking up through the COAS in the overhead windows. They'll just tweak burn us to keep *Friendship* centered in the middle of the COAS."

The general rule of thumb is that the shuttle pilots set up a closing rate of one foot per second for every 1,000 feet. At 500 feet the speed would be closed down to 0.5 feet per second; at 200 feet down to 0.2 feet per second, and so on. Rendezvous must be done slowly. The slow closing rate is essential so that many jets don't have to be fired when the shuttle gets in close. From 100 to 120 feet out, exhaust plumes from the jets can affect the stability of the station. That's when low "Z" thrusters are fired out fore and aft, from the nose and the tail of the shuttle and not directly toward the station.

Within 100 feet of *Friendship*, the commander of *Discovery* would have to move in very slowly, without the use of radar, since radar is not reliable within the 100-foot range.

The rookies listened intently as Chico outlined the fascinating details of the rendezvous. He finished by explaining to them the importance of the position of the sun.

The correct position of the sun is essential in a successful rendezvous. Initially, during the close-in rendezvous, the pilots have the sun to their backs. They want to have the final closure done before orbiting around the Earth when they have to face the sun. They also want to be docked before they go back on the night side where they might have trouble seeing the station. There would be about 50 minutes of sunlight to work with before they would drop into the dark side for the other 40 minutes of their orbit.

"Question?" one of the rookies broke in.

"Shoot."

"Why is it a 50/40 break on daylight to dark? I don't understand."

"Fair question," Chico responded. "Just remember that we're at a considerable altitude above the Earth up here. So, even though we may fly over part of the Earth that is still in darkness below us, we're up above so we can see the sun more quickly as we come around from the dark side. And the same is true as we go around on the other side of the world. We stay in sunlight a bit longer up here at this height. So an orbit isn't half daylight and half nighttime. We have 50-minute days and 40-minute nights."

Meanwhile, up on the flight deck they were very carefully skinning the eyelid of a bear. Smokey had driven down this track many times before. Now this white-haired veteran crept slowly up on the cylindrical target ahead using his manual DAP system. Then, with a light touch on his translation hand controller he danced along the X, Y, and Z axis coaxing *Discovery* forward. He did not want to disturb *Friendship* any more than necessary, because that would affect the experiments being conducted on her internal racks. Even creeping up on her at 0.1 feet per second, he would create a 500-pound force for a period of one second at the moment of contact.

"Contact."

"Roger, *Discovery.* Confirm capture lock. Welcome to the Tiltin' Hilton."

"Thanks, troops. Good to be here again. Hope your bridge game has improved. We'll be powering down here and let you fellows fly this old ship for a couple days, what say?"

"Sounds good, Smokey. Come on in, coffee's on."

"I was hoping you'd say that. Three-thirty came awfully early this morning. Request permission to come aboard, *Friendship.*"

"Permission granted, *Discovery.*"

Half turning in his seat, Smokey smiled at his passengers, who had now gathered behind him to gawk at the giant station.

"Ah, we've been given permission to board, troops, but while Bob gets this bird into its nesting mode, I think I should cover a few things. I know you've probably heard all of this before during training, but up here we play hardball, and I'm required to remind you of that before you board. I know that your focus the past few months has been almost entirely on your experiments. And I know that all three of you are pretty capable scientists in your own right. But up here there's only one boss and we better settle that right now. Do you all understand? I want you to leave your egos here in *Discovery,* and I'll take them back down with me and return them to you when your tour is finished."

No one spoke, but all nodded agreement. This was not the kind of welcome speech they had expected, but it was just the right tone for the serious business ahead.

"As you know, Bob and I are going to leave two of our crew here with you to help assemble the new panels on the solar dynamic system. Chico and Ann have both been here before, and you could do a hell of a lot worse than look to them for advice. If something gets in your craw, don't let it fester; that can be fatal for the entire crew up here. Just talk it out. You kids got that?"

Discovery was joined safely to *Friendship* at the primary berthing port on the node at the end of Habitation Module One. The shuttle docking mechanism was on a telescoping tunnel at the front of the payload bay. Fully extended this tunnel would reach 27 inches above the orbiter's outer mold line. It mated with the outboard port. There were six such ports on each of the nodes. They permitted docking by the orbiters or future expansion into adjacent modules. Four of the ports were around the side walls at 90-degree angles; the other two were at each end. The hatch through which the crew would now pass was 50 inches in diameter. This permitted the easy loading and unloading of experiments and supplies.

Each module was connected by these 12-foot-diameter nodes. There were also several 12-foot-long tunnels about 6½ feet in diameter to provide access to all areas.

The habitation modules had two 20-inch-diameter windows in them and four 10-inch windows. The laboratory modules had just four 10-inch viewing ports. The windows were designed so that two people could look out of them at a time. They not only held in the cabin pressure of 14.7 pounds per square inch but they also provided protection from meteoroids and outside debris. They could be replaced while in orbit if they became damaged. The air gap between the two panes of each window was vented out into space to prevent moisture from building up.

Two of the rookies gestured that Mary Two Hawks should enter the station first. She pushed off into the connecting node, and Wayne Morrison quickly followed, anxious for the adventure to begin. As they quickly disappeared through another hatchway, Billy Wong, the third rookie, held firmly to the handrail in *Discovery*'s tunnel, peering into the now-empty connecting node. It was far larger than he remembered from the ground mockup.

Mary Two Hawks, Wayne Morrison, and Billy Wong were the first doctoral candidates to visit *Friendship*. They were each researching in different fields, but they had been able to make a case for the value of a stay at the space station for their individual theses. Their 30 days on *Friendship* would immeasurably enhance their research.

Half entering the node, but still holding onto the handrail with one hand, Billy glanced up to the top of the room and gasped in surprise. Through the open hatchway in the ceiling was the waiting universe. He paused, mouth open, as he realized what he was looking at. It was the

It takes practice to learn to move effortlessly from one module to another through the connecting nodes in the absence of gravity.

observation cupola, the small attached roomlet that projected out into space so that the crew could have a 360-degree view outside the main structure of the station.

"I can take you up there later so you can do all the gawking you want, Billy," Chico broke into his thoughts. "But right now we'd better get on in and say hello to the rest of the crew."

"Sorry, it just all sort of overwhelmed me, and I'm afraid I don't feel very good. I hope I don't barf."

"Well, then, we'd better get on in and you can stay still for a while. It's best if you don't move around too much at first. I know it can do darn funny things with your insides when you don't have gravity keeping all of your organs in place any more."

Billy could see into the softly lighted module ahead through the open hatchway. It seemed a quiet, restful area, rather like looking down a peaceful dormitory hallway. Reaching out tentatively he pulled himself through the hatch, and before he knew what was happening, his body was moving into the module. It was then that he made his first error in judgment aboard *Friendship*.

His momentum had carried him into one of the side walls and, stopping himself with his hands, he instinctively pushed off with his legs against the floor and side wall. They had warned them against doing this, and they had even practiced how to move around in weightlessness. But in his excitement Billy forgot his training, and now he flew quickly across the hallway into the opposite wall.

"Ouch, damn it, I jammed my finger," he swore as he ricocheted off the bulkhead.

"Careful there, youngster," Chico warned. "Don't push off too hard. If you just use your fingers and sort of pull yourself along, I think you'll find that it's a lot more stable and a heck of a lot safer."

"Rats, I'm sorry," Billy said dejectedly, "I've only been here for a couple of minutes and already I've screwed up."

"Forget it, Billy, if that's the worst thing you do while you're here you've got nothing to worry about. Why don't you let me lead the way down into the wardroom and you can see how I do it." And with that, Chico moved off effortlessly down the length of the module.

Ahead of them, Mary Two Hawks and Wayne Morrison had passed through the dividing partition into the other room at the end of the hall. For the first time, they were aware of the odor of the module. It was not the sterile plastic and electrical equipment smell they had gotten used to in their ground training. It was sweet, but with an off-center smell they couldn't quite identify. In reality, the environmental control system con-

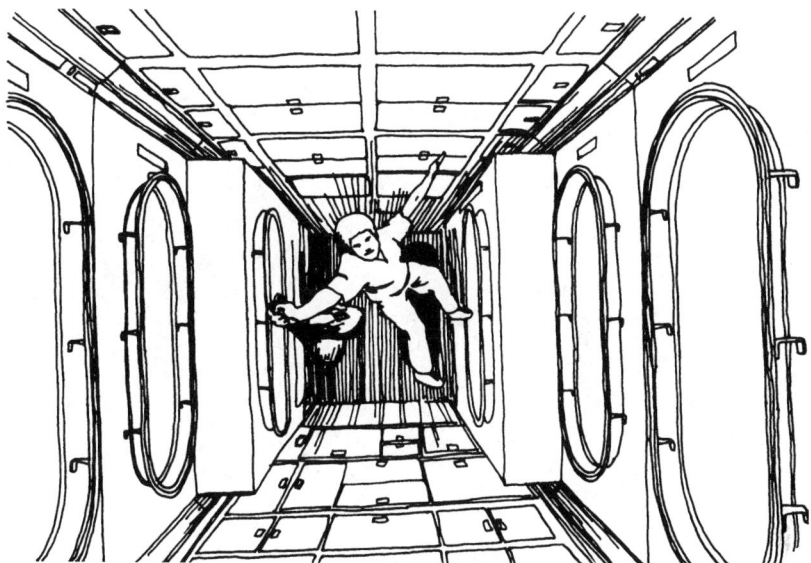

This is the first view newcomers to *Friendship* get as they leave the docked space shuttle. Formerly the sleep area in Habitation Module One, this portion of the space station has now been converted to commander's office, doctor's office, and library.

tinually struggled to keep ahead of the ever-changing blend of odors that was created in this tightly closed room.

"You all look like you don't quite know what to make of the place." Chico laughed as he followed them into the combination kitchen, meeting room, and exercise area. "If it's the unusual aroma, don't worry, you'll get used to it. From what Smokey has told me, they got some very strange smells when they first started flying the shuttle, too. Matter of fact, they discovered that they couldn't even store apples and bananas in the same place. For some reason the combination of the two fragrances drove everyone nuts. They finally tried separating them and that worked fine, but now, because of the longer tours, they don't even bother to bring the bananas along. So, if you want fresh fruit with your meals or for snacks, I'm afraid you'll have to stick to things like apples that don't spoil as quickly.

"Well, listen, I'm going to go back and help the guys unload some of the smaller things we brought along. Why don't you just make yourselves at home. I'll be back in a few minutes."

"Thanks for your help," Billy said sheepishly.

"Listen, just forget it and enjoy the place. See you later."

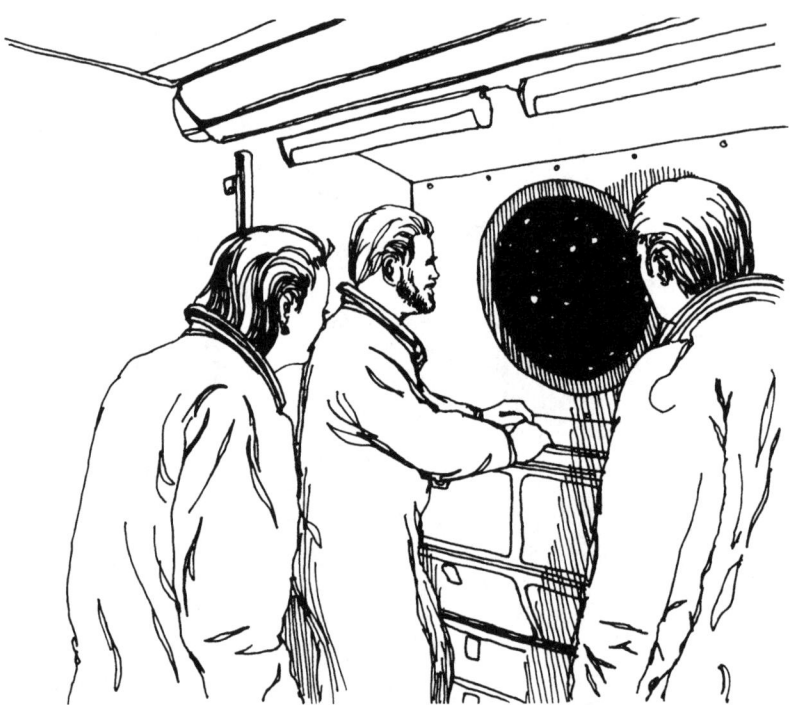

The view out the window of the space station is a powerful magnet to all who serve aboard her. Every scene is different, and with the changing orbital path, weather changes, and day/ night sunlight changes, it is a constant source of entertainment and wonder.

The three rookies swam awkwardly over to the two large portholes on the right side of the module. The view was straight back down at Earth and they could see the meandering spines of a mountain range below them. It blended into plateaus and finally down into the sea. The Earth seemed to be rotating slowly below them rather than them speeding along at over 17,000 miles an hour.

"I wonder how long it's going to take us to figure out what we're looking at from up here?" Wayne said, almost to himself.

"Maybe we never will," Mary answered quietly.

2

Orientation_____

After several minutes the rookies' reverie was broken by a voice behind them.

"Welcome aboard! It's good to have you with us."

They turned clumsily toward the voice, aware of not having this weightlessness business down pat yet. It was Station Commander Stu Robinson, whom they had all met the previous spring during training in Houston. Stu's specialty was astrophysics, in addition to welcoming space rookies on board. There was a fine mind behind that easy smile and boyish looks, and Stu took to command easily.

"Well, you finally made it up here. Good to have you with us. The head's over there if any of you have to visit before we get started. Try to be neat."

Soon the important things were out of the way, also rather awkwardly, and they all gathered in the galley area.

"We might as well get started on your orientation while Smokey and Bob power down the bird. I've got a list of things I'm supposed to cover with you in the next few days, but feel free to break in at any time if you have a question. You're going to be here for 30 days and we all want to be sure you're comfortable, have a productive work visit, and get to enjoy the sights.

"Are you starting to feel green yet?" Stu asked quietly as the rookies' attention returned once more to another new view out of the seductive portholes.

"A little," Billy said, smiling nervously.

"Well, don't let it bother you; just go ahead and enjoy the misery. Almost all of us feel that way the first day or so. It kind of goes with the territory. My advice would be to skip lunch right now and see how you feel a bit later. Then try something light like soda crackers and see if you keep them down all right. Let's just spend some time in here for a while so that you don't have to move around too much. Keep your body in the 1-G trainer heads-up position as much as possible; we find that that helps keep the stomach settled the first couple of days.

"We really didn't get to know one another too well when we first met, so why don't we spend some time right now and get better acquainted, sort of like the first day at summer camp. I'll start, okay?"

The three rookies smiled self-consciously as they tried to slide their feet into the restraining loops on the floor.

"Wait. Let me unfold the table and then you'll have something you can hold onto as well as using those foot restraints." Stu unlatched a boxlike cluster of panels from underneath the counter running along the side of the module, then swung the sections of the cluster apart until they formed a row of individual desktops. Moving down the counter, he did the same to a second wing of the stored tablelike unit. Finally he hooked the two sections together into a semicircle that adjoined the counter.

"Now, that should help. We usually keep this thing in the stowed position and out of the way except for meals or meetings. I know you saw the mockup in Huntsville during your training, but here's how the computer screen works. Just pull up on the top of the thing like this and snap it into position. The keyboard simply folds down against the top of your work area like this. Then you can reach down under the unit and snap your desktop up into the locked position. There, now you have work space or eating table."

"That's really neat," Billy said with a smile.

"Let's see," Stu said. "Where to start? Well, I was raised on a farm in northeastern Kansas until my folks lost it back in the eighties and we had to move in with my mother's family in Kansas City. After high school I moved back out to the University of Kansas and got a degree in aeronautical engineering. I also earned a commission in the Air Force through the ROTC program and entered flying school after I graduated. One thing led to another and before I knew it I was out at Edwards flying new aircraft.

Space station commander Stu Robinson welcomes the three rookies aboard the orbiting "think tank" as they gather for the first time in the "family room" of the space station, the wardroom.

Finally, when the call came for more astronauts I applied, and on the third try I was lucky enough to get accepted.

"When they expanded *Friendship* in 1998 I was on the second orbiter flight in the left seat, and I liked it so much up here that I put in for a future tour of duty. Nothing much happened until I earned a master's in astronomy at Rice during night school, and that just sort of added the icing to my resume. I made it up in 2003 as part of the science crew updating the *Hubble*, and here I am again, only this time as the boss-man."

"How many shuttle flights did you make?"

"Three, not including the one I rode up here on, of course; that one was just as a passenger. Now, how about you? Mary?"

"Well, my name is Mary Two Hawks, that you already know. I was born in Oklahoma, but my mother moved back east to North Carolina after my father died in an auto accident. She finally got a job as a park ranger and worked in Flat Rock at Carl Sandburg's home, Connemara Farm. Actually, I spent almost as much time there as my mother did, especially in the summertime when I helped out in the goat barn. Mrs. Sandburg was pretty well known for her Chikaming dairy goats, and I helped keep the barn cleaned out and fresh hay in the feeders, but mostly I played with the barn kittens."

"How old were you then?" Wayne asked.

"Oh, I was in grade school by then. Probably nine or ten. I used to take a peanut butter and jelly sandwich, some cookies, and a jar of goat's milk up to the top of Big Glassy Mountain there on the farm. It really isn't a mountain, but everyone calls it that. I used to climb up there a lot. It's where Mr. Sandburg used to go all the time to think. Up on the top there's a big, smooth, rocky area and you can look out over the Blue Ridge Mountains.

"One night I talked my mother into camping out up there with me. It was the same summer that *Friendship* here was first started. I remember wishing we could see it from where we were, and we did see something moving high up against the stars. I know now that it couldn't have been *Friendship* because of her orbit, but back then I thought I was looking at the new space station, and I dreamed of someday coming up here and here I am."

"It must have been the Russians' station that you saw," Stu said. "It has a much higher inclination orbit than we do. And it was already up by the time they started work on this place."

"Well, I suppose it was, but at the time I thought I was looking at *Friendship*. Anyhow, a couple of years after that my mother remarried and we moved to Houston. After I graduated from high school I enrolled at the University of Houston, and I've sort of been there ever since."

Mary Two Hawks, a Native American, is working on her thesis on the psychological stresses of space travel on family relationships. She is from Houston, Texas.

"How did you get interested in your thesis topic? Space stress, isn't it?" Billy asked.

"Yes, actually the formal title is 'Stress Factors in Long-Duration Flight as They Relate to Familial Relationships.' I was working at the Johnson Space Center during the summers as an intern. During my second summer, I worked in Building 4 where the astronaut offices are located. It didn't take me long to see that some of them had some really serious marital problems.

"As I dug into it, I found out that NASA had always spent a great deal of time and money training the flight crews themselves, but they'd done virtually nothing about the families that stayed down on the ground. They were all just supposed to be good soldiers and put on a good show. Just think of the stresses that deep space flights will put on marriages and parent-children relationships.

"But don't get me started on it, I could talk for hours. My mentor, who encouraged me to pursue the thesis, was Dr. B. J. Bluth at headquarters. She's been the resident expert in this area for years.

"Well, that's about it for me. I'm single, but not really looking too hard yet; there's just too much I want to do with my career before I get ready to settle down. Oh, and I'm 25 years old."

"Thanks, Mary," Stu said. "I envy you your time at Connemara Farm. Sandburg is a favorite of mine. Those must have been great years. Now, who wants to bare their soul next? Billy?"

"Okay. I'm Billy Wong, I'm Chinese-Canadian. My grandparents came to Vancouver, British Columbia, where my mother and father were both born first-generation. They moved out east to Chapleau when my father got a job there in the paper mill. So I grew up there in Ontario. No, I never played hockey; sorry to disappoint you. But I did do a lot of river guiding in high school and college. As a matter of fact, that's really how I got interested in astronomy and ended up here."

Stu laughed. "Okay, I give up. That one you're going to have to explain."

"Well, one of the places I always took my walleye fishermen was downriver out past the rapids. There was Crown land on both sides of the river, so it was nice and wild. The tourists always like that, and on overnight trips we always used to camp with them at the base of the rapids. You could get some really good fishing downriver from there, see a few moose, sometimes an eagle, and then camp on the high rocky ground south of the rapids. It's really a neat spot."

"Sounds like my kind of place," Wayne said.

"It's really a fine place to grow up. And wildlife management is what first took me down to the States to Michigan. Not that we didn't have some

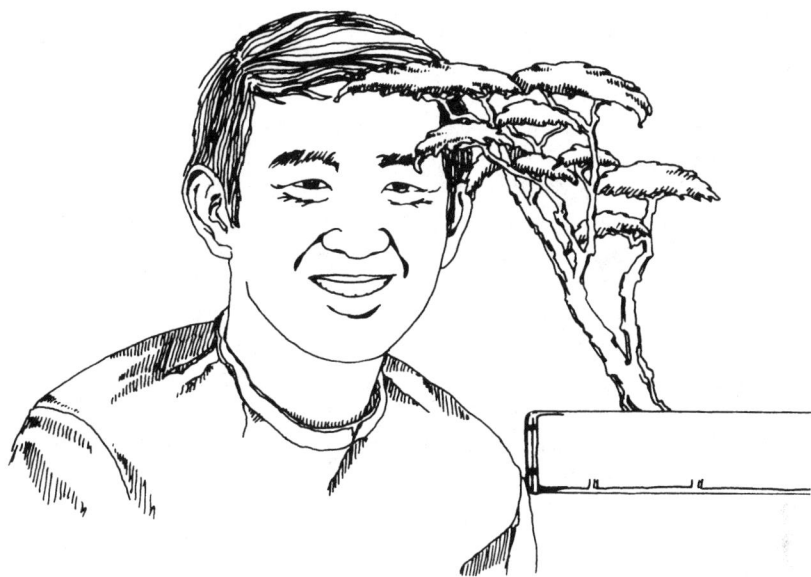

Billy Wong, of Chinese-Canadian heritage, is aboard *Friendship* to further his studies on the syncopation patterns of black holes. His home is Chapleau, Ontario.

good schools in Canada, but one of our summer people taught at Michigan State and talked me into enrolling there. So that explains my rather unusual combination: bachelor's degree in wildlife management and doctoral work in astronomy and astrophysics."

Stu looked puzzled. "I understand the wildlife management part, but what does that have to do with astrophysics?"

Billy giggled nervously. "Sorry, I guess I left something out. The summer I graduated from MSU I came home for the season to guide, and that's when I met Dr. Brown—Sam Brown, as he insists I call him. He and several other people from the astronomy and space sciences department at Cornell came up fishing for a couple of weeks. They were headed for Wawa on Lake Superior, but their car broke down outside of town, and so they ended up fishing for walleyes instead of lake trout. Somebody recommended that they hire me to guide them and that's how we met.

"Fishing was lousy for the first couple of days, but camping out with them at the rapids was unbelievable. At night they introduced me to the universe and my options just exploded. I never dreamed there was so much out there, or so darn many questions still needing answers. I drove them nuts with questions. Really, I probably ruined their vacations I had them talking shop so much.

"We finally got into some nice messes of walleyes, though. And in the fall I enrolled at Cornell. I met Carl Sagan there my first semester, and he convinced me to switch my field to astronomy."

"So, after you got your master's degree at Cornell you decided to go for the union card, right?" Mary asked.

"Yes, by then the Ph.D. wasn't so frightening. I'd learned to look at the top of the mountain from time to time, just to keep it in sight, and to concentrate on each individual step. Next thing you know I was accepted for this flight to gather information for my thesis, and here I am."

"And so I hear you're now working on black holes instead of fishing holes," Stu said.

Billy grinned. "Yes. My thesis is titled 'Energy Absorption and Syncopation Patterns on the Thresholds of Black Holes.'"

"Gulp. That one you're going to have to explain to me in more detail another time." Stu laughed and turned to Wayne. "Well, Wayne, I guess you're next. What brings you up here with us?"

"Well, actually, I came up to outer space to help me see more clearly into inner space."

"Say what?" Billy said, relieved that his show and tell moment had passed.

"You know, my cell research."

"Before you get into that, Wayne, how about where you were born, things like that."

"If you can believe it, I was born in Sydney, Australia, and raised in Gnaw Bone, Indiana."

"You've got to be kidding," Stu said.

"No, seriously. My folks were appearing Down Under with the symphony at the Opera House there, and when it was time, it was time. God doesn't care where he plops us down; that's up to us to sort out later. Luckily Sydney was the last stop on the tour, so we just stayed there until my mother and I were strong enough to travel, and then they brought me back home to Brown County."

"Gnaw Bone, are you serious?" Billy asked.

"If you think that one's funny, just up the road north of Nashville there's a Beanblossom, Indiana. My first true love was from Beanblossom."

"That sounds like it was out in the sticks like where I grew up," Billy said. "What were a couple of symphonic musicians doing living there?"

"They both taught at the Indiana University School of Music, which was only about fifteen or twenty miles west of Gnaw Bone. Every summer they toured with the Indianapolis Symphony."

"So is that where you went to school, too," Stu asked, "at Indiana?"

"Just the first two years." Wayne paused and then added, "My father suddenly died of stomach cancer, and I grew up damn fast." The silent attention of the others encouraged him to go on.

"It was just such a shock. He really hadn't seemed to be too sick, just losing weight was all. I mean, he didn't even have any pain until the last couple of days. The day before he died the doctor called my mother and me into his office to talk to us. He didn't even offer us a seat. He just took us into one of his examining rooms and told us flat out that Dad only had a day or two left.

Wayne Morrison, of Gnaw Bone, Indiana, by way of Sydney, Australia, is studying a possible breakthrough in cancer research dealing with the induced aging of cancer cells. One of the major reasons for *Friendship*'s existence is its major contributions in research on cancer, diabetes, and kidney disease.

"The next night Dad took a turn for the worse and had to ask for a shot of pain killer. It was the first and last one he ever asked for. I sat there on the edge of the bed and held his hand while they gave it to him. I remember a pigeon landed on his windowsill and we talked about it. Later I learned that some people in the old days considered a dove's appearance a sign that death is near. But at the time I felt that the Holy Spirit was comforting us all. Really, a dove and a pigeon look an awful lot alike.

"I sat up with him all night while my mother tried to get some sleep. I must have dozed off just before dawn, because suddenly I woke with a start when the birds started to sing and the sky grew light. I checked to be sure Dad's chest was moving, that he was still breathing. I was relieved that he had made it through the night. But then, all of a sudden, his breathing changed, and as I pressed the call button for the nurse he took three quick breaths, and then there was the death rattle. I was glad his suffering was over, but I felt guilty at the same time for thinking it."

Wayne glanced at the others, suddenly embarrassed by having said so much. Stu reached over and squeezed his shoulder. No one said anything for several minutes. Then Wayne spoke again.

"It happened so quickly that I didn't know what to do. I mean, they don't give you lessons on that, and they really should. The birds were still singing when the nurse finally came in. I looked away when she shone the flashlight in his eyes. I looked to the windowsill expecting the pigeon to be sitting there, but it was empty."

To break the tension that had settled over the wardroom, Stu asked, "Is that when you got interested in medical research, Wayne?"

Wayne took a deep breath and rubbed the back of his neck. "Yes, it was as if my father's death gave a purpose to my life. Up until then I thought only of what would make me happy as far as a goal in life was concerned. I had changed my major twice. I was really all tied up in myself. But when Dad died it was as if he gave me one final gift, and I knew that I had to try to find out why he and so many other people die each day of this thing we call cancer. That's when I transferred up to the IUPU campus."

"IUPU?"

"Sorry, that's the joint Indiana and Purdue campus on the southwest side of Indianapolis—the medical school is there. And so I became a medical researcher. When I discovered their flow cytometer, it opened up a whole new world for me in my research on oncogenes. My experiments up here in electrophoresis are really a joint project between IU, Eli Lilly and Company, and the National Cancer Institute. It's all proprietary, of course, but if what I test up here works out as we think it will, then Lilly will produce enough of it for our human clinical tests. So far it's worked very well with rats."

"Exactly what is it, Wayne? Can you talk about it at all?" Mary asked.

"Yes, a little bit. This new drug is capable of triggering aging in cancer cells. You may not be aware of it, but cancer cells don't seem to have the biological clock that other cells do in the human body. They can live forever. They just don't die like normal cells do, they never age as they should. They're missing a gene that we call a sengene. We're hoping to shortstop cancers from starting. All of us at the Hoosier Oncology Group are keeping our fingers crossed.

"We're working with other cancer research centers around the country in trying to understand the 40 or so oncogenes that are in each human cell. So far we feel we have a good handle on only about a half dozen. That leaves one heck of a lot of unanswered questions down inside human cells. I'll be working on my little piece of the puzzle while I'm here with you. Maybe it will open some more doors for us. Thank goodness *Friendship* is up here."

"Okay, I feel like I know you all a little better," Stu said. "Now, for some basic information.

"You all know the drill, but let's go over safety one more time. The station is assembled in a raft configuration permitting two exits from each hab and lab module. So if you need to get out, you can scoot through either end. Be sure to pull the hatch closed behind you to seal the area; the hatches can be opened from either side. If it becomes necessary to dump the atmosphere overboard in event of a chemical contamination or a fire, that will not necessarily disable that module.

"The few times we've had things like this in the past we've been able to use the sensors and vent and replace fairly quickly. It's better to be safe than sorry, people, so don't let things get out of hand, just punch the panic button. As you know, they're located at each work station.

"Now while the orbiter is still here it's powered down but still configured for an emergency escape and undocking at any time, so that's a good thing to remember before the crew leaves in a couple of days.

"In the event of a 'hole in cabin' emergency from a meteoroid hit, you should also evacuate that module when the alarm goes off and we'll do a damage assessment. We can always do an EVA if we have to. We can locate the hole from outside and do a hot patch from there with our portable toolkit, but in most cases all we do is follow the oxygen loss from inside. Then it's a relatively simple matter to unhook the rack, rotate it up out of the way, and seal the hole from in here.

"Now, if we have to do a rapid EVA, we can egress pretty quickly with our new hard suits. As you may know, the EMU service station is set up now to reservice the EMUs 20 times a week rather than 10 as in the old days when *Friendship* was first sent up. So that's a big help. Any questions so far?"

"Have we ever had a fire in one of these living areas?"

"Not that I know of, at least none has ever been recorded by the crews on the data base. Most of our problems in that regard happen in the manufacturing and lab areas. We have a lot of exotic ovens and equipment in those modules, and sometimes we do screw up and have to scramble to keep things under control. But we've been very lucky so far, no major structural damage.

"Graffiti. Let's not forget the important things like that. We used to discourage graffiti, but even the *Apollo* crews couldn't resist leaving their signatures on their command modules.

"So rather than fight it, NASA in all its wisdom has decreed that the wall in your private sleeping area may be used for one signature by each crew member abroad this here bus. Please act accordingly. One to a customer.

"I'd better mention lighting. As you already know, but maybe you've forgotten, we have a sunrise and a sunset about every 90 minutes. That's 16 'days' every 24 hours. Now our computer, the infamous HAL, figures all of this out for us every week and we post a new crew schedule. It doesn't do to have one of you working up a sweat on a bicycle when one or the other of us is trying to do a solar event observation through the H-alpha instrument. The bike shakes the hell out of everything.

"But I started to tell you about the lights. We have night-light route indicators and switch illuminators in all areas, so don't panic the first couple of days before you get used to the constant changes. It kind of sneaks up on you at first.

"Next item. You're required to take one shower per week and use the wet wipes every day. Period. No exceptions. I don't care how fragrant you like to get. Up here we can't put up with it. And please observe all water management techniques when you do use that facility. Let's not rust shut from the inside.

"Flatulence. 'Farts' to the great unwashed. It's a real problem up here because of what weightlessness does to the natural body gases inside your system. We all have it, but let's all try to be considerate. If you get a bad attack, retire to another module for a while.

"Vitamins. You were provided with an ample supply. Take them. Colds make us all cranky and out of sorts. You'll also find that your sinuses don't drain properly up here and colds can be damned uncomfortable, so a word to the wise.

"You will, underline *will*, observe quiet hours in Hab Two. While you are in here you can play cards, watch movies from the fresh supply you brought up with you, listen to cassettes, or whatever. Even sing in the shower. But please, we are all working different shifts and we'll all be sleeping, or trying to sleep, at different times. There will be no exceptions to this. If you use your VCR or audio station in your sleep area, you must use your earphones, or keep the sound level turned way down.

"Speaking of shifts. We are on a five-day, nine-hour shift on this tour of duty. You will get two days off per week if you care to take them. One day

Friendship Timeline—Mission Specialist

Work Day

Sleep
Personal hygiene
Urination/defecation
Dressing/undressing
Meal preparation (breakfast)
Eating
Meal cleanup
Planning and scheduling
Materials processing experiments
Payload support
Urination/defecation
Hand/face cleansing
Planning and scheduling
Meal preparation (lunch)
Eating
Meal cleanup
Materials processing experiments
Urination/defecation
Materials processing experiments
Meetings and teleconferences
Exercise
Hand/face cleansing
Meal preparation (dinner)
Eating
Meal cleanup
Medical care
Urination/defecation
Private recreation and leisure
Small-group recreation and leisure
Dressing/undressing
Full-body cleansing
Personal hygiene
Private recreation and leisure
Urination/defecation
Sleep

Off-Duty Day

Sleep
Meal preparation (breakfast)
Eating
Meal cleanup
Hand/face cleansing
Sleep
Urination/defecation
Dressing/undressing
Exercise
Full-body cleansing
Personal hygiene
Meal preparation (lunch)
Eating
Meal cleanup
Clothing maintenance
Private recreation and leisure
Training
Urination/defecation
Hand/face cleansing
Meal preparation (dinner)
Eating
Meal cleanup
Personal hygiene
Dressing/undressing
Sleep

off is mandatory. Two are advisable. Now I know there's a great temptation to work all the time on your pet experiments, but it's just too easy to burn out up here, and you will need to take time off to sleep in and look out the window.

"You also need to talk to your families every week. We can't force you to do that, but it's strongly recommended for your sake as well as theirs. Your comm times are marked on your weekly schedule that HAL makes out every Saturday night.

"Speaking of time, all on-board clocks are set to Houston time. So if you still have your watches set on Cape time, this would be a good point to change them to avoid confusion.

"One last thing, and then we'll take a break. You know that you brought up a supply of beer and wine. This is for our attitude adjustment hour after each day's shift before dinner. Please. The rule is just one beer per crew member per day. We can't run out and pick up an extra six-pack up here, so please don't abuse the privilege. One cold one before supper really hits the spot. Enough said."

"Sitting" at the table was taking the newcomers a little getting used to. Seats were not necessary, of course. NASA had learned its lesson on that score with the design of the old *Skylab,* the first space station. There had originally been a seat in *Skylab* at the solar console, but the first crew removed it. The console turned out to be too low, and the astronauts' stomach muscles hurt all the time from the constant bending over.

"Move nice and slow, but unhook your feet now and let's cover the rest of the equipment in this end of the module. You've probably already noticed the television screen up here between the two portholes. You can either watch live broadcasts with your meals or during off-duty hours if the area's not in use for meetings, or you can watch the videotapes stored down here in these areas.

"Be sure to close the storage areas tightly whenever you get anything out of them anywhere on board. It can be a real mess if you leave them partway open. Things have a way of floating away and disappearing. We find more unusual things stuck up on the air returns.

"This is the work station over here. It's actually just one of several on board, but since this is where I usually hang out I do most of my work from here. I'll show you later some of the things we can do with it. Incidentally, if you ever want to find me, you passed my office on your way in. I'll cover that end of the module with you in a little while.

"Back here is the galley, but before I get into that, maybe I'd better show you where the galley storage is. This is your first real lesson in orientation. Swing your bodies around like I'm doing. There, now the floor has become a wall, how about that. Everybody's stomach still okay?"

The galley storage compartments ran all the way down to the sides of the module. There were 14 compartments, varying in depth from 36 to 40 inches. In the same area were drawers for the data management system. The trash collector and housekeeping supplies were in compartments at one end.

The CO_2 reductor and concentrators and the O_2 generator were stored in larger compartments. In others were the contaminant monitoring and control unit, a portable H_2O dispenser, and the atmosphere microbial removal control equipment. A pyro control box, portable processors, and condensate tanks were stored in still other compartments.

"Now, moving very slowly," Stu said, "let's reorient ourselves back again to the galley wall. Slowly, now."

It was no use; Billy lost the battle with his stomach. Luckily he had stuck a barf bag in his pocket earlier.

"Better now?" Stu asked gently. "Don't be embarrassed. It happens to the best of us. Maybe for your mental health we ought to go down to the other end of the module now and leave this galley area for later. Probably do you good not to be thinking about food for a while until your stomach settles down. Let's cover the exercise area over here and then go back into the other end."

Billy smiled weakly.

"You need to take it easy on the liquids, too, for a while," Stu told them as they left the galley area. "When we first started this business we thought dehydration was going to be a problem, and then we learned that just the opposite was causing us a lot of unnecessary sickness.

"Well, this is our exercise area, as you can see from the treadmill and the exercise bike. After your workout you can use the shower and dressing room back on the other side of the partition. When we assembled *Friendship*, this was one of the first modules we brought up. That's why we still dock with it. It's sort of a tradition. Back then we did all of our living, sleeping, cooking, and exercise in here. But now we've converted it for day use and moved the sleep compartments out into the quiet modules. We did leave the shower, dressing room, and john in it, though. Let's go back there now and I'll show you what we have back in that end."

The partition around which the group now floated was really little more than a heavily padded canvas, off-white in color. It simply acted as a visual separator from the wardroom area, the office, and the medical sections of the module.

"Okay, now this area over here is our health management facility. That's always been here. Here is the combination washer and dryer, and over here is the shower and dressing room. On the other side of the hall is the john, as you've already found out. I'll get into those more when we get over into the sleep module.

"Now, from here on back down toward the node you entered through,

Regularly scheduled exercise is an important part of each tour aboard the space station. The equipment is positioned so that the crew members can see the exciting views out the windows as they work out.

we've made a lot of changes. As you can see these side compartments can be pulled out into the hallway to give us more room. That gives us an area inside the compartments of a little over four feet in depth. So you can see that here in this office—this is mine by the way—the room is four feet by seven feet. Incidentally, the ceiling is also seven feet high out in the aisle. Now, across the hall from my office we've fixed up the doctor's office the same size. He's over in the JEM today helping Mitsue and Shoji with a life sciences experiment. You'll meet him later.

"Next to the doctor's office is our Ames Library. You can use it for a quiet spot to find some solitude when it's free. We also use it for meetings and conferences. Now, across the aisle from it is our pharmacy and my storage area. Each of them is a rack wide. Now, don't forget, all of these units, these racks, are removable so that we can get back behind them for repairs and the like. It's a real sweet design job. Now, let's go back down this way and I'll fill you in quickly on the HMF.

"We have the usual storage bins on the ceiling and floor just as we did at the other end of the module. You can look at them later if you'd like to wander around. Housekeeping supplies, ORU manuals, light bulbs, flashlights and batteries, cameras, film, computer paper, all of that kind of stuff is kept back here.

"Now here next to the HMF most of the ceiling and floor storage is medical supplies. Let's see, up there is the back-up defibrillator and the IV systems, also the laboratory supplies and the anesthesia equipment. This compartment has microbiology kits and physiological monitoring supplies."

Billy touched Stu on the arm and pointed toward the ceiling. "What," he asked quietly, "are those? Are they really what I think they are?"

"I'm afraid so, Billy. They're just what the label says—body bags. That's one of the hard, cold realities up here. We have to be prepared to deal with death. I know it's an overworked expression, but it's a very hostile environment up here. And we're working in it every day. Death does not ignore noble undertakings. But let's look over here for a minute. The equipment in this area is what we call the HMF, the health management facility. It takes up an area some six by seven feet overall.

"As you can see, we have a very full complement of analytical medical equipment here. It's amazing the amount of sophisticated equipment they squeezed into this wall. We've got a centrifuge, an incubator, and a digital imaging light microscope. There's equipment over here for urinalysis and a digital X-ray unit. Also, an osmometer, a hematology analyzer, and a whole blood analyzer. Over here is a metabolic/respiratory testing system, a transcutaneous gas analyzer, and a 12-lead ECG/arrhythmia monitor for our coronary research. Here's the ventilator, the defibrillator, and a pulse oximeter.

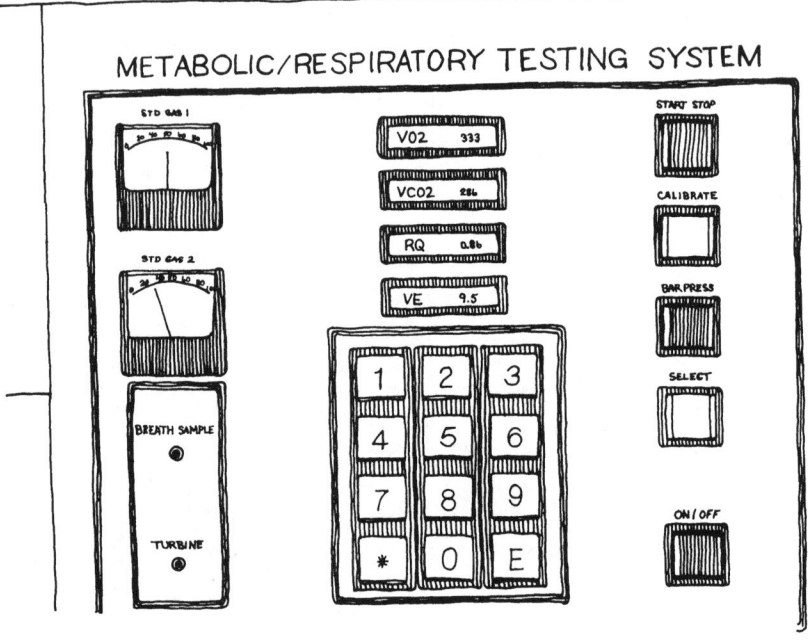

A variety of routine tests and emergency procedures can be performed at the health management facility with the sophisticated equipment on board.

"It's a very complete facility, as you can see. But you'll see more of it as time goes by. I understand the three of you have volunteered for the new series of baseline tests the good doctor is doing. So, how's the stomach now, Billy?"

"Okay, I guess."

"Think you can handle the galley now?"

"I sure hope so. Let's give it a try."

"Good. Incidentally, down here in the floor is our fecal gas processor and our hygiene water processor. The urine pretreat and processor is down in this compartment, too. Solving the problem of recycling urine for drinking water was a real breakthrough for us back in the eighties. Not too many people ever thought we'd be able to do it for one thing, and then get our crews to actually drink it. But you tried it in your training and I know you can't tell the difference between it and fresh water. Let's go on back in here and cover the galley now."

"Can I ask a question?" Mary said.

"Of course. What is it?"

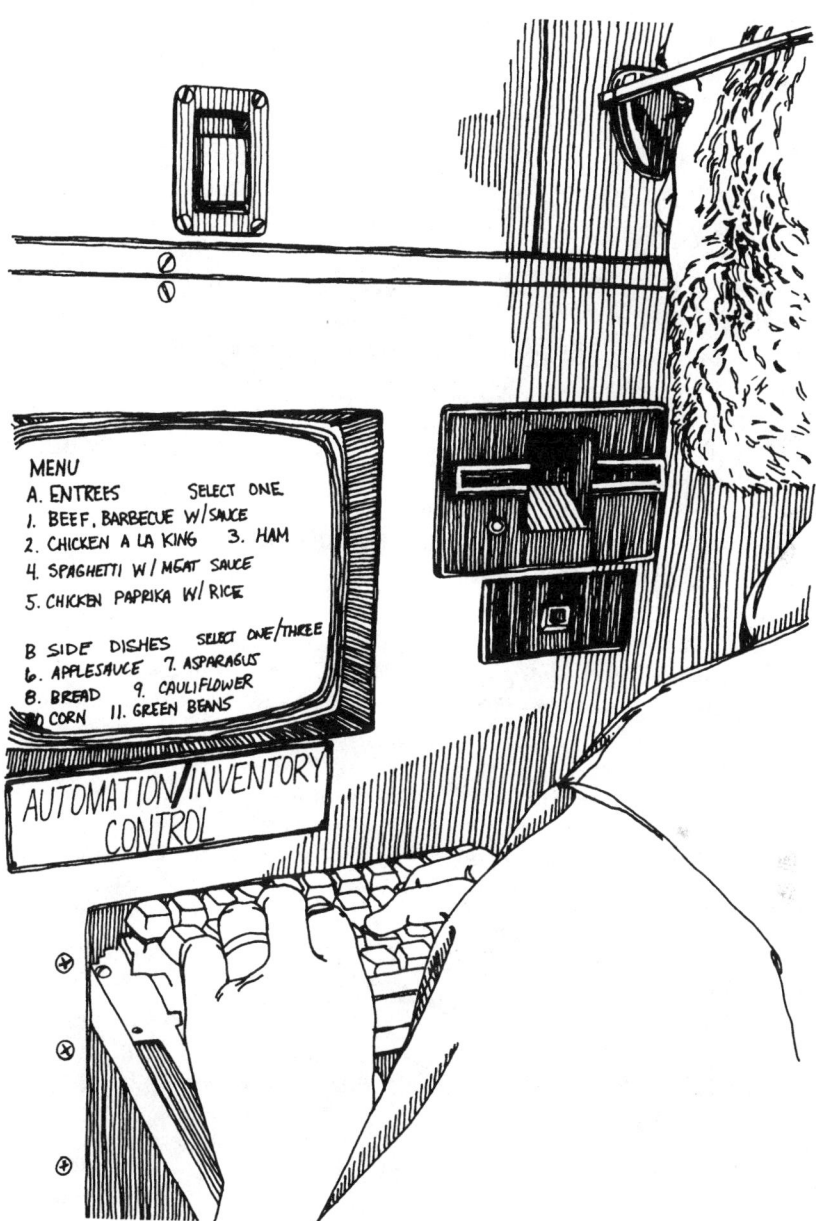

MENU
A. ENTREES   SELECT ONE
1. BEEF, BARBECUE W/SAUCE
2. CHICKEN A LA KING   3. HAM
4. SPAGHETTI W/MEAT SAUCE
5. CHICKEN PAPRIKA W/RICE

B SIDE DISHES   SELECT ONE/THREE
6. APPLESAUCE   7. ASPARAGUS
8. BREAD   9. CAULIFLOWER
10. CORN   11. GREEN BEANS

AUTOMATION/INVENTORY CONTROL

A crew member asks the computer, HAL, what is on the menu for the evening meal. All food items are entered into the computer, and as they are removed for use, a scanner automatically deducts them from the "pantry." The same type of system has been used by supermarket managers since way back in the 1980s.

"Well, I know that sometimes you all take turns cooking and all, but do you do that every day, or do you just eat out of the prepackaged food, too?"

"That sort of depends on the makeup of the crew during a tour. Most of us are usually too busy to do much cooking, but there's always somebody on board who likes to do some of it just to relax. So most of our meals are the prepackaged ones they send up from the Cape. Here, I'll show you.

"This is our automation/inventory control screen. This helps us keep track of the inventory we have in the freezers and refrigerator over here. We can keep a 14-day supply of meals right here in the galley. The rest of the food we keep down in the logistics module. We have freezers down there, too. Let me call up a day's menu for you.

"Here you can see that for an entree you have your choice of beef barbecue with sauce, chicken à la king, ham, or spaghetti with meat sauce. Side dishes are applesauce, asparagus, bread, cauliflower, corn, green beans, broccoli, Italian vegetables, potatoes au gratin, or rice. Then for dessert, you can have brownies, fruitcake, peaches, pears, chocolate pudding, or strawberries. Now, all you do to see what the beverages are is hit the return button over here.

"Up here are the convection microwave ovens with their keypad controls. These drawers hold ambient food that doesn't have to be frozen or refrigerated. Down here are the dishwasher and a work surface that you can pull out. Here are dry wipes and dry trash bags, and over here are the wet wipes and biocide wipes. The trash compactor is down here, and the inserts are stored next to it. Snacks are kept in here. Now, remember, be sure to close all of these storage compartments after you get into them.

"The fresh vegetables that you brought up with you are a real important touch to have up here. Somehow the aromas of freshly peeled onions and crushed cloves of garlic can have an awful lot to do with one's peace of mind. Well, I think that's a pretty fast once-over on this module. Let's take it nice and slow and go over to the sleep module, and I'll show you your rooms."

3

The Quiet Zone_____

Stu's guided tour of the space station had brought the three rookies into one of the tunnels that, with the nodes, linked the space station together. This design had been adopted in order to conserve work space in the modules.

If the modules of *Friendship* had been locked together directly, in what was known as the racetrack configuration, a large portion of each module would have had to be used for docking ports and all the associated lines that interface for power, oxygen, drinking water, and fiber optics data buses. The node and tunnel design makes optimum use of module space while allowing several modules to be linked together. Engineers call it the raft pattern.

The nodes are the size of a small room, 12 feet in diameter. The interconnecting aluminum tunnels are 12 feet long and 6.7 feet in diameter, wide enough for two crew members to pass each other or for equipment to be moved from one module to another. They are lighted and have handrails and circulating fans.

As Stu directed the others through the tunnel he pointed out that that particular tunnel was known as the Hall of Fame.

"As you can see, there's a mission patch attached to the wall here from every crew that helped build *Friendship* and that has lived in it since. Let's head on through now into the next module, what we call our quiet zone."

The others followed rather awkwardly, a laugh or two punctuating the air.

"Well, I see everyone made it through safely," Stu teased. "Don't worry, you'll get used to making your way around this place in a day or two. I know it's tricky at first. The important thing is not to be in a hurry. Newcomers always push off too quickly and sail across the cabin too fast. You've probably already done it, haven't you? Just take it easy. Do it all in slow motion and you'll soon get the hang of it.

"Now I thought this might be a good time to give you some background into how our 'house' up here got this way. As you know, we have three habitation modules now. That gives us the capability to have 24 people on board at a time.

"We do all of our sleeping in this end. That's why we like to keep the noise down as much as possible. I've split you up so that each of you will be in a different hab mod. We figured that you'd adjust quicker that way.

"I don't know if they covered the scenario in your training on how *Friendship* was built. Did they? No? Well, let me take a couple of minutes to bring you up to speed."

Stu told the rookies how NASA thought it would take 14 shuttle flights to put up the space station when they decided on the dual-keel configuration. When the number of flights was jumped to many more than that, they decided to put up a big part of the station with unmanned heavy-lift vehicles and to stretch the construction out into several phases.

The two vertical keels of the dual-keel configuration are each 361 feet long—just a little longer than a football field. They are connected at top and bottom by crossbeams, called booms, which are 146 feet long—14 feet less than the width of a football field. The 503-foot-wide transverse boom running across the center makes it look like a square figure eight.

The station was taken up in chunks and put together one step at a time. The pieces had to fit inside the orbiters or on the heavy-lift rockets. After being assembled at an altitude of 220 nautical miles, the station was boosted up to 250 miles.

The keels and booms used to construct the framework are actually trusses and are 16.4-foot cubes. Struts are made up of 0.060-inch-thick tubes made from a graphite-epoxy composite. They have an outer diameter of 2.12 inches, about the size of a tennis ball.

The trusses not only provide the structural foundation but also permit utility lines and payloads to be attached to them and to run along their length. They give the station its stiffness for better altitude control. In addition, they provide a roadbed for the track that carries the mobile remote manipulator system—the Canada-built arms that lift things out of the shuttle bays and move payloads around the structure of the station. The trusses provide a good backup structure for load stresses and can be repaired without destroying the integrity of the entire structure.

The struts were folded up inside the shuttle cargo bay and deployed little by little from the orbiter once they arrived at the proper spot to ensure an orbit of 28.5 degrees. This position was chosen because 28.5 degrees is the inclination to which the maximum payload can be delivered by the shuttle, and because the majority of U.S. missions can be accommodated at 28.5 degrees.

The 16.4-foot cubes are laid out so that nine of them make up the crossbeams at top and bottom. The main keels are nineteen cubes long, with eight cubes above the center boom and ten below it. The center boom is seventeen cubes wide. Attached to this basic "flying figure eight" are solar panels, radiators, servicing and storage hangars, solar dynamic power units with 50-foot parabolic reflectors on them, and experimental packages to continuously look at both Earth below and the universe above. Also attached to the structure are all the habitation and laboratory modules, the logistics module that is changed every 90 days, and the modules of the European Space Agency and Japan.

After describing how *Friendship* was built, Stu continued. "Well, while we're right here at this end of the module, I might as well fill you in on it. The sleep compartments are all at the other end, on the other side of that divider wall. We'll get to those in a minute or two. At this end as you can see, we have sort of a mini-galley area. This is just for snacks and your early-morning coffee. There isn't even a dishwasher here, so everything you use in here should be disposed of over in this bin.

"Now, we've tried to keep these hab modules as quiet as possible, but we know that you won't be sleeping all the time when you're in here during your scheduled sleep periods or on your days off. So, that's why we have this galley and study area as well as the television set over there. But this TV doesn't have a speaker system. If you want to use it you'll have to use the headsets. They're stored in that drawer under the set.

"There's a pretty good selection of movies as well as educational subjects in this storage area. If you bring any tapes over here from the Ames Library in Mod One, you're honor-bound to return them. Gotta keep peace in the family, you know.

"As I said, this whole end of the module is for your convenience for your off-duty hours, and it's designed so that you can use it and not disturb the people trying to sleep on the other side of the partition. That's why there isn't any exercise equipment in here, or a washer-dryer. You'll have to use the equipment in the other module for those things."

"I see there's more than one john," Billy said.

"Right. We tried to keep standing in line to a minimum, so there are two commodes in each of the sleep modules. But there's only one shower and dressing area. NASA figured that most people would change in their own compartments with the exception of their weekly shower, so they didn't duplicate that area. Let's look into the head for a minute so I can show you where everything is.

"There's not room for anybody but me in here as you can see, so you people just stay out there and I'll answer your questions. Let's start with the commode first. This is completely different from the one that was on the shuttle in the early days. That thing was a disaster. We don't have any sewer lines. That could have been a real mess if we'd had a clog develop along the line somewhere. And transferring waste out into space was just too damn tricky and dangerous to try to do.

"We have a pretty simple chemical system that collects the waste, deodorizes it, and draws off the fluid that's in the solid waste. The urine is recycled, as you know, for part of our on-board supply; that's why we all use this separate device here. I know they covered all of that in your training. If you have any problems using it, don't be bashful about asking one of us to help refresh your memory on it. You guys can ask me, and Mary, you can ask Edie or one of the other women."

"I have a question." Wayne interrupted. "You said you separate the solids from the liquids in the fecal matter. Then what happens? I know I should remember, but I forgot."

"Okay. Let me go through it one time. The fecal bags are stored over here in this compartment. If you need one and they're all gone, we keep a 90-day supply right over there in that compartment on the floor. See it?"

"This one here?"

"Right. There are packages of them stored in there as well as extra wet wipes, toilet tissue, Kleenex, and sanitary napkins. All of the replaceable items for this compartment are stored in there.

"The procedure is that you take this porous bag and fit it around the unit down here like this. Then, after you sit down and strap yourself in, you simply turn on this air flow switch. The air keeps everything settled down in the bag. When you're done, you seal the bag, thusly, insert it into this second bag, and put it into the heat-vacuum chamber down there in the floor. That's where it's dried.

"Every week or so we transfer the packages down into the logistics module for return to Earth for analysis. Our medical people use the solid waste for ongoing experiments about our diets and body chemistry. Now each of these bags is treated with a chemical to take care of odors and the growth of bacteria, so you don't have to worry about putting in tablets for that purpose like the guys did in the early days of the space program.

"Just one word of caution. This system is not foolproof. And you're responsible for cleaning up your own mess if anything escapes until you learn to use the equipment properly. I don't mean to be insensitive, but it can get tricky when you get diarrhea. So take a good look around the place when you're done, and make sure you don't leave anything floating around. If you do, you can use the wet wipes to clean up. They're stored in this compartment next to the lighted mirror."

"So we can clean up and shave in here too," Billy said.

"Just as long as there isn't anyone waiting in line. As I said, that's the nice thing about having two commodes in this module, but we still get backups every now and then at our peak hours in the morning and at meal times.

"Well, I think that pretty well covers waste management, as the engineers call it. All the storage areas are labeled as you can see, so there's no guesswork. Just be sure to close the doors tightly when you take anything out. And turn off the mirror lights when you're done. Let's move across the hall to the shower and dressing room.

"Maybe I should remind you of some of the things that your body is going to be doing at this stage, before you stumble in here some morning and see the stranger in the mirror. For one thing, you know about the body fluid shift that you're already experiencing as everything seems to move up into your chest. Your intestines just aren't pulled down by gravity anymore.

"Your face will also begin to look different. Loose skin on your face floats up and your face gets a puffy look. You'll also probably have a full feeling in your head from all the congestion. Your sinuses just won't drain right up here, and even the blood vessels in your neck and scalp will bulge. Now a lot of this goes away after a couple of days, but your face will still look different to you most of the time you're up here, so don't panic at the new face in the mirror."

"That's why they fitted us for different clothing for our flight bags, too, wasn't it?" Wayne asked.

"Yes. It's adjustable so that you can still be comfortable as your body changes size while you're up here. That's something NASA learned the hard way. A lot of astronauts were pretty uncomfortable in their space suits in the early days. The suits were tailor-made for them while they were down on the ground, but in space our height increases up to a couple of inches and a lot of us lose four inches or so around the waistline.

"Your legs will get pretty thin, too. Matter of fact, people call 'em

The "dressing room" just outside the shower. On the far wall, an important feature is the full-length mirror, put there at the suggestion of psychologists so that crew members can maintain a good personal image.

chicken legs. That's also caused by the fluid shifts upward and the loss of muscle tone. Your height changes because you don't have gravity pulling down on you any more, and the discs between your vertebrae expand. So your spine gets longer and straighter up here, too. Incidentally, has anyone ever explained to you why you only brought a 14-day supply of clothing along with you?"

"No, I don't think so," Mary answered.

"Well, it doesn't save so much for a 30-day visit like you'll be making, but just for the eight people that live in this module a 14-day supply of clothing weighs 240 pounds and takes up 84 cubic feet of storage space. A 90-day supply would weigh 1,520 pounds and take up 517 cubic feet of space. And at $1,000 a pound to get it up here, we'd rather bring up more scientific equipment than clean underwear. That's why we have a washer and dryer in the 'day' module."

"You said that we can only shower once a week. Is that right?" Mary asked.

"Afraid so. We do encourage you, though, to take a sponge bath every day. Let me get in here and you guys can look over my shoulder.

"This is the dressing area. As you can see, once you close the curtain you'll have plenty of privacy. The full-length mirror was added at the suggestion of psychologists. They figured that it might help the morale and mental health to see yourself like you do down below.

"Washcloths and towels are stored in here. Just throw them in your laundry bag after you shower and do them with the weekly laundry. Don't forget that we like to have you do your laundry together as much as possible to save on the water. That's why we asked you to put your initials on everything during preflight.

"Now you probably know that you only get a gallon of water for your shower. It's not exactly like at home, but at least you can wet, lather, and rinse your whole body. Just be sure to dry off well and use the water vacuum to clean the residue off the walls before you lower the wall. We also suggest that you use the hair dryer so that you don't get chilled. You can come back in here later and see where everything is stored. Let's get Wayne settled in his room now; it's down this way."

"You say Mary and I are in other modules?" Billy asked.

"Yes, we thought it would be better that way. Now, Wayne, this is your room. You can put your name tag up here over the door after you get settled in. That'll help everyone get to know it quicker.

"Now, as you can see, we keep the lighting in this end of the module rather subdued. All the fixtures are one rack wide so that they can be removed when we have to change out a room or a rack full of equipment. Here, let me show you– they snap out just like this. Well, anyway, the book says it's easy. This one is stuck for some reason. I'd better not force it. Sometimes we get a little warpage during the light and dark sides of our orbits.

"Again, there's only room for one of us in these units. So Wayne, since this is your space, why don't you get in and I'll show you what's what from out here. Can the rest of you see okay?

"Your room is 78 inches tall and 53 inches wide in the extended position. Theoretically, you can shove the walls back 14 inches when you're not using the space, but in reality we all leave them extended out into the hallway unless we're moving equipment around in the aisle.

"When you're in there, Wayne, you have 31 inches between the back wall and your equipment wall. It may not look like much at first, but you'll find that it really is enough, especially when you figure that you can use all the volume. Down below, most of the airspace in a room goes to waste. You can get real cozy in here if you like to burrow into a corner."

"So this is my home away from home, finally, after all this time day-dreaming about it," Wayne said wistfully.

"This is it all right," Stu said. "Now, down here is where you can store your clothes, your day-to-day stuff. There's also a compartment out here in the hall ceiling where you can keep your other things that you brought up. We'll unload all that stuff later this afternoon, and you can unpack and get settled in. You can store your garment bags out here, too.

"Now, over here are your personal restraint harnesses that you can use while you're in here. We suggest you keep them stored in here when you're not using them. Up here you can store your videotapes, earphones, and computer paper. Your emergency oxygen mask is in here; we'll run through its use again later this afternoon. Your tissue and dry wipes are in here."

"This monitor looks different than the ones we trained on in the mockup," Mary said.

"Oh, really? You must have been using the ones that they're going to be adding to the new modules. We don't have those up here yet. But basically I think you'll find that these are not too much different.

"The monitor up here shows you the status of the critical station systems, such as electrical, air, pressure, and temperature. Your temperature controls are in easy reach here, and you can direct the air flow just like you do in an aircraft. Your TV screen works only with the headsets as I mentioned before. And your computer screen is here. The keyboard releases with this latch so that you can fold it up out of the way when you're not using it.

"Well, that about covers that. Your sleeping bag unhooks like this if you ever have to take it down. Oh, and you can put your personal photos

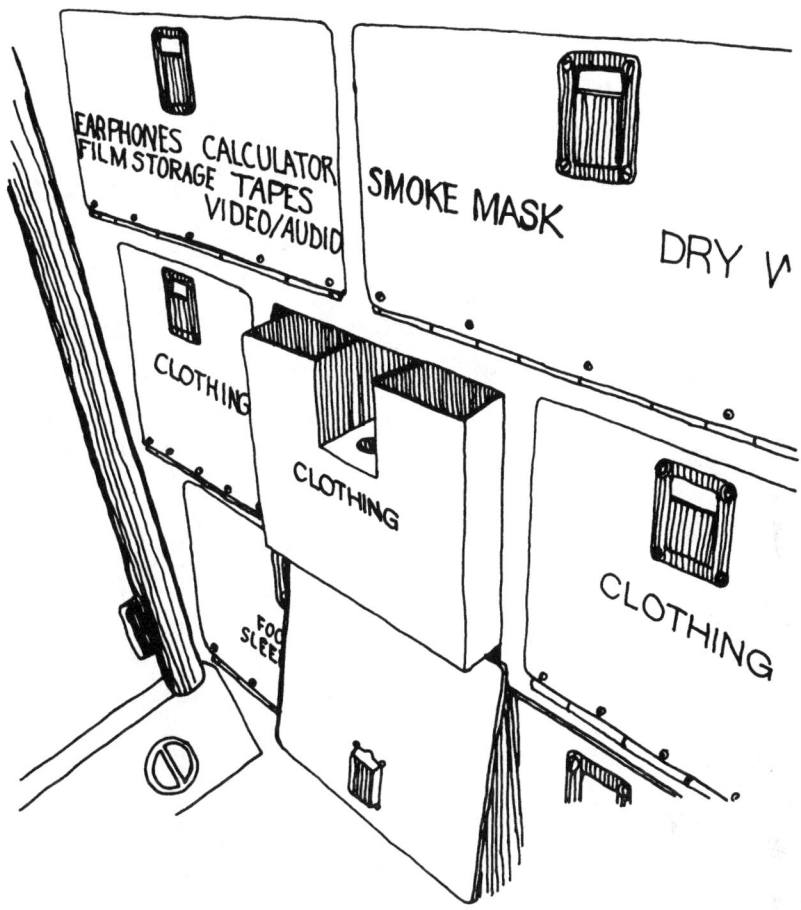

Some of the storage facilities in the personal sleep compartments.

in these little sleeves here. We tried to keep the Velcro to a minimum in these cubicles. That used to really bug the guys in the old *Skylab*. Whenever anybody got up in the middle of the night to go to the bathroom, they woke everyone else up stripping their Velcro apart. So you'll see that we don't have it in here on high-use items.

"One other thing that you should know about. You'll notice that you have a cassette tape capability here in your room. One of the things that people up here miss the most about life on Earth is the sounds. So the married folks often bring up cassettes they've taped of the sounds around their homes, of their children and grandchildren. We've got some tapes

Billy Wong at the work station in his sleep compartment.

you might like to listen to after you begin to get a little homesick. There's also a complete library of music tapes.

"We encourage you to keep a diary. It's amazing what a journal can do to help you fight depression, even if you just write a couple of sentences a day covering what you did or how you feel. A lot of us go through a period of depression after the first thrill of being up here wears off. A journal habit can help smooth over something like that.

"You should also be prepared for trouble sleeping at first until you get used to weightlessness and not having the pressure of your head on a pillow. This thing creaks and groans all day long, too, as it goes from hot to cold. Anyway, don't try to fight your feelings; just accept them and work through them. We all go through it whether we like to admit it or not. There are no supermen or superwomen up here, just us ordinary folks who miss our teddy bears.

"Once you've been here a couple of days," Stu continued, "you'll notice there's very little difference between our duty and off-duty clothing. Funny T-shirts are okay. But you must wear clothing appropriate for the work area you are in. And on times like a national holiday when we televise our life aboard *Friendship* for the public, we like to put our best foot forward. No off-color T-shirts on those days. Off-duty you have your choice as far as clothing is concerned."

Stu pointed out the hand-held Halon fire extinguishers and the smoke, oxygen, and heat detection system. The smoke sensors trigger a master alarm that sets off flashing lights and a siren in case of a fire. A message is displayed on the CRTs for the benefit of those who are at work stations and in the sleep areas.

A cabin pressurization valve can be used to drop the internal pressure in each module so that within five minutes a fire will not have any oxygen to keep it going. The system is designed so that the depressurization rate will not be harmful and so that the crew will still have time to evacuate the module after opening the valve.

Part of the system also monitors water pumps and avionics fans that cool the equipment. Caution signals are set off if problems develop in those areas as well. (Avionics are the electronic systems and instruments on board used to monitor and control things.)

Each module is equipped with a thermal system to take care of the heat created by body metabolism, on-board experiments and manufacturing, and the hot glare from the sun during every orbit. Outside temperatures actually swing from 150 degrees above to 20 below zero on every orbit.

All this heat needs to be dissipated so that it does not cause condensation or discomfort inside the modules. The heat is carried away from each

A *Friendship* crew member catching some Zs in the sleeping bag in his personal sleep compartment. Eye masks are optional.

area to the radiators out on the boom. The radiators—one on each end of the boom—wind up and down on every orbit to protect their fluid joints from the cold. They can be rotated at orbital rate to keep their edges pointed toward the sun.

"Question?" Wayne asked.

"Fire away," Stu responded with a smile.

"Do we have adequate protection against meteoroids?"

"Good question. Yes, each module has meteoroid bumpers built around the outside of it with about a two-inch space between it and the skin of the module. That's where we have our thermal insulation.

"We do get quite a bit of scouring, though, from star dust and atomic oxygen. Matter of fact, we have to recoat the radiators periodically because of that. You know we're moving along at a pretty good clip up here, and incoming dust sort of has a sandblasting effect on the whole station. If you did an EVA, you'd notice it on the leading edges of everything out there. No problem as long as we keep up with it on sensitive areas. Just like keeping up with the barnacles at sea.

"When the shuttle is docked with the station as it is now, its bottom surface is facing the line of travel, so it provides a substantial buffer for the station from the stray meteoroid. During the days of Apollo moon flights, the command module windows were examined after flight and a total of ten meteoroid impact craters were found on the windows of those spacecraft. Five of them were on the *Apollo 7* ship alone.

"It's been estimated that 200 million meteors enter the Earth's atmosphere every day, but most of them are no larger than a grain of sand. During the year Earth meets great swarms of meteors as they orbit around the sun. The Earth goes through this stream every November and we have the annual Leonid meteor shower. There are ten such close encounters each year with different groups of meteors. And whither Earth goest, so goes *Friendship* station."

"Stu, Code Six!" Stu's lecture was interrupted by a voice coming over the intercom system.

"Got you, Beth. We're in Hab Two."

"We need you for a while over here in the ARF. Can you break away?"

"Can do. What's up?"

"Daisy is giving me fits. She won't let me give her the injection today. She's all arms and legs. I need help holding her down. It shouldn't take long."

"I'm on my way. Ah, you kids will have to entertain yourselves for a while. I'll be right back; they need me over in the animal research facility." Stu pushed off toward the node, twisting his body to adjust to the new vertical, and disappeared into the tunnel.

"What do we do now? I'm afraid to touch anything," Mary said.

"Well," Wayne answered, "we have to start sometime. Just don't flip any switches. I don't want to end up landing in Shaker Heights."

"I think I'll try the hand washer," Billy said. "I didn't have time to do that after I went to the john. My mom would kill me if she found out."

The hand washbasin was rather like the isolation containers in the lunar receiving lab. Those containers had ports through which the scientists could reach to handle the moon rocks with special gloves. The Lobund Germ Free Laboratory at the University of Notre Dame developed this technique back in the late 1940s. However, in this hand wash facility there were no attached sterile gloves. Billy just reached through the ports; the plastic sleeves kept stray droplets of water from escaping out into the cabin. A stream of air flowing over his hands pulled the water out into the wastewater storage and processing tanks for reuse.

"There, that wasn't so difficult."

"Here's one of those wind-up shavers," Wayne joked. "Anybody want to knock off a five o'clock shadow? We have a safety razor and shaving cream, too."

"No, but I'll want to shave my legs later, silly," Mary laughed.

4

Animal House_____

Mary Two Hawks, Billy Wong, and Wayne Morrison were wide awake early next morning even though they hadn't slept much the night before in their new quarters. As Stu had warned, it wasn't easy sleeping with your head barely touching the pillow. And the unfamiliar sounds, the groans of the space station, the constant awareness of activity made sleep a challenge.

None of them complained, even though they all agreed it seemed as though they had been trying to sleep sitting up. But the excitement of all the new experiences and sensations concealed any fatigue. They were eager to hear about the Animal Research Facility.

"Welcome to Animal House. I'm Beth Tippett."

The rookies had not met Dr. Tippett before, although her reputation was well known in the academic circles they traveled. Her work in mammalian development and reproduction in weightlessness had earned her the lead life sciences position on *Friendship*. And she enjoyed showing off her domain to newcomers.

"We deal in reality here, people," Beth went on. "No denying that we are studying our own small moment of evolution, but we study what will happen off Mother Earth, not on it. Our main function is to observe the physiological effects of zero-g as well as the overall role of gravity on living systems."

"Question," Wayne interrupted. "Why is this wall up in this module?" He was looking at the wall running across the middle of the module they had just entered.

"Do any of you remember the *Spacelab 3* flight way back in the eighties?"

"Is that the one they called NASA's Ark?" Billy volunteered.

"Yes, that's it. Do you remember the problems they had with it?"

The rookies had blank looks.

"Well, on that flight with the new Spacelab module in the orbiter's payload bay they had two squirrel monkeys and twenty-four rats on board. They were testing out the air-flow containment racks as well as a lot of our early procedures. Unfortunately, the food tray change-out system didn't work properly, and bits of food and feces formed a cloud of dust in the cabin of the shuttle and drove the crew bananas. You can imagine how well they enjoyed their meals with monkey and rat doo floating all over the place.

"Anyway, that's one reason we have our zoo walled off. Another is to help control the noise level in the station for both the animals and the crew. Environmental control is another. Incidentally, we use an activated charcoal system for odor control. So this end of the module is the ARF, our Animal Research Facility, and the other side is full of racks for our life sciences experiments."

"Are these just regular pet store animals that you use up here in your research?" Mary asked.

"No, as a matter of fact they're not. To protect our crew's health we can't bring animals on board that may be disease carriers. And we also have to get good, reliable data for the medical research we're doing. So most of our rodents are gnotobiotic. We get them from commercial breeders."

"I'm not sure I understand," Billy put in.

"Gnotobiotic animals are born by Caesarean section and raised in a germ-free environment. We double check them for every known pathogenic agent just to make sure. As a matter of fact, they do need certain microbes in their systems to survive, and so we give them a cocktail of bacteria right after birth. Now there are no such things as gnotobiotic primates. They get the gut bugs when they nurse. But we test them very laboriously to make sure that they are free of pathogenic agents."

"How many animals do you have on board right now?"

"Let's see. There's Daisy and Arabella II and . . . but maybe I'd better list them by groups. One squirrel monkey, forty-eight rats, about a dozen pocket mice, two garter snakes, a colony of fruit flies, six cross spiders, a school of mummichog minnows, and a very animated tray of African clawed frogs.

"Docked to the right of the node down there is the garden module. We have some honey bees and other assorted insects there that I can show you later. We keep them over there to pollinate the plants we're studying. But I'll cover that later.

"You know, I really can't tell you how many living things we have in Animal House and here in the life sciences racks. We do basic cell research in life sciences, and I guess we just don't know exactly how far you can go yet in breaking a cell down and learning about its minute parts. Is each cell a living thing? Is each part of one cell a living thing? You know a cell contains about a quarter of a million protein molecules alone, and the scale that we do much of our research on is mind-boggling.

"In all of this we're trying to understand and define microgravity effects on genetics, reproduction, embryological development, maturation, and population dynamics. And in order to do so we have to have long-range studies not only on animals like us and like Daisy but also on many generations of space-born plant and animal species.

"It has always seemed so easy to us throughout history to relax and assume that everything there is to know has already been discovered. But, thank goodness, every generation or so we get a gentle boot in the rump.

"You know, there are more mysteries to all of this than there are answers. Several years back some of our NASA scientists found one of the simplest forms of bacterial enzymes ever known. The little bacterium that this enzyme lives in thrives in a saltwater environment and may be a leftover from early evolution.

"So what we have here at *Friendship*, in essence, is research in cells going back to perhaps the first moments of the creation of life on Earth, while at the same time our Hubble telescope out the porthole there is reaching out 14 billion light years. And with it we are seeing the light that left those galaxies back when the universe was born. In here we are watching parts of cells that go back to our own beginnings. The continuum of time becomes a reality here.

"Working in this part of *Friendship* makes you feel humble and realize your own short span on Earth. It's what drives us and makes us all want to accomplish just as much as we can on every duty shift. We feel that the secrets of the universe and of life are being revealed to us daily. I wonder if some greater force knows the ecstasy of learning about our creation one tiny bit at a time. Maybe that's why we didn't understand it all from the beginning."

Stu had been quietly hanging in the background, letting Beth have center stage with the newcomers. "Tell them about the insulin pump, Beth," he now interjected.

"Yes, the insulin pump. Another one of the things we're working on is to reduce the size of the implantable insulin pumps that NASA animal research came up with back in the eighties. That saved an awful lot of diabetics their daily sessions with the needle. We're now working on a smaller version.

"But even more exciting is the beta cell work we're doing. We're almost there with pure beta cells. Once we can make them next door in the lab in large enough volumes, we'll be able to inject them into a diabetic's liver and cure diabetes. There are about 3 million people who would get down on their knees for that breakthrough. I'll tell you more about our pharmaceutical work another time.

"You may not know, but six months after *Spacelab 3* landed in 1985, researchers announced that the 24 rats that had flown on board had been dissected and studied thoroughly. They uncovered a serious roadblock to long future spaceflights. The rats showed a dramatic reduction in the release of growth hormones during the flight. One solution for ongoing missions to Mars and the outer planets might be extra doses of growth hormones during flight.

"One of the things we're doing with Daisy is working to determine the correct dosage of these growth hormones. So the work we're doing here in the ARF could very well benefit our forthcoming joint mission to Mars with the Soviets.

"Our space station life sciences program," Beth continued, "includes both basic science and applied science. Our applied science goal is to make sure that our crews are healthy and happy during long space flights so that they can survive and be productive on the scientific missions we send them on for a permanent presence in space. Basically, we're trying to learn all we can about the origin, evolution, and distribution of life throughout the universe. Also how life forms are affected by the harsh conditions found in space.

"Our goal is not only to understand us but also everything that affects us. And that's no small task. The challenges aboard *Friendship* are staggering, to say the least. But so are the potential rewards for the betterment of mankind. The eradication of disease, pestilence, famine, and war are all within our reach.

"There's really a lot more I could tell you about our research in this area," Beth said, "that is, if you want to take the time, Stu?"

"Be my guest. That's what an orientation period is for, isn't it? Don't worry, I'll wake 'em up if they fall asleep."

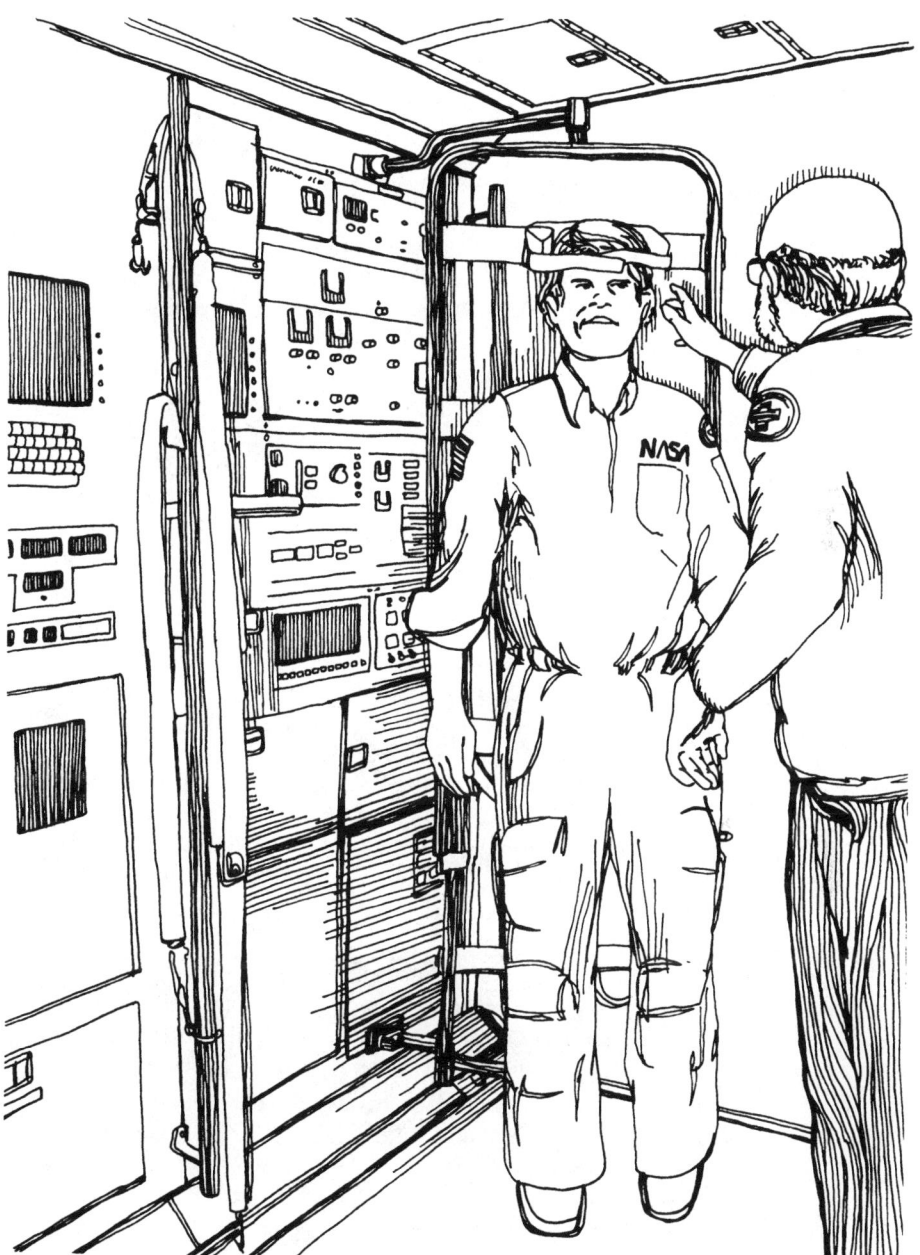

For some medical tests, and in emergency situations, there is from time to time a need to immobilize a crew member. It is done in this tablelike fixture, taking up a minimum amount of local-vertical floor space.

"Okay. Well, what we're really searching for is how to develop the health maintenance programs that will become routine on long-duration space flights well into the next millennium. Understanding radiation, radioactive particles, particularly high-energy particles, are all on the agenda. We're also studying the psychology of space flight. We need to be sure we understand all the stresses on crews who are working for long periods of time in remote environments like this, where people are confined and have no chance to escape. On submarines and Antarctic bases they have similar problems. Did you know that on just two cruises of one Polaris submarine, 7 out of 137 men on board had some form of psychiatric disorder? So anxiety and depression are two very real problems on our shopping list that we must solve."

Beth then summarized their work on six other systems of the body. In the areas of mineral and hormonal balance they were studying bones, feces, and urine to further understand bone development and diet, nutrition, metabolism, and the digestive tract. On *Skylab* flights the astronauts' bones were actually dissolving during the flight. They lost up to 2 percent of their minerals per month. A 20 percent loss could cause serious bone failure, particularly to the back and leg bones, once they return to the gravity of Earth. So Beth's team had been working to find a solution.

Another area they were monitoring carefully was that of hematology and immunology. They were studying red blood cells—their metabolism, production, and destruction—and the body's defenses against disease enzymes, serums, meiosis, and bone marrow. The benchline *Skylab* crews suffered an average 10 percent loss of total red cell mass. It was thought at the time that red cell mass would recover after an initial decline, but similar research by the long Russian flights did not bear this out.

The cardiovascular system was being studied to determine the effects of gravity on circulation, blood pressure, and the dynamics of the heart—hydrostatics, acoustics, and thermal dynamics. The heart does not have to work as hard to pump blood in weightlessness and so it loses some of its size. During the *Apollo 15* flight to the Hadley Plain area of the moon, both Dave Scott and Jim Irwin suffered cardiac arrhythmias. After returning to Earth, Irwin suffered numerous heart attacks over the years. In both cases a loss of body potassium was the culprit. Both the *Apollo 16* and *17* crews were put on a high potassium diet 72 hours before launch, continuing until 72 hours after they landed back on Earth. This helped a great deal, although they, too, had some isolated premature heartbeats on their flights.

Energy expenditure was another major area of study. Breathing, metabolism, exercise effects, heart rates, calories, and the oxygen/CO_2 rela-

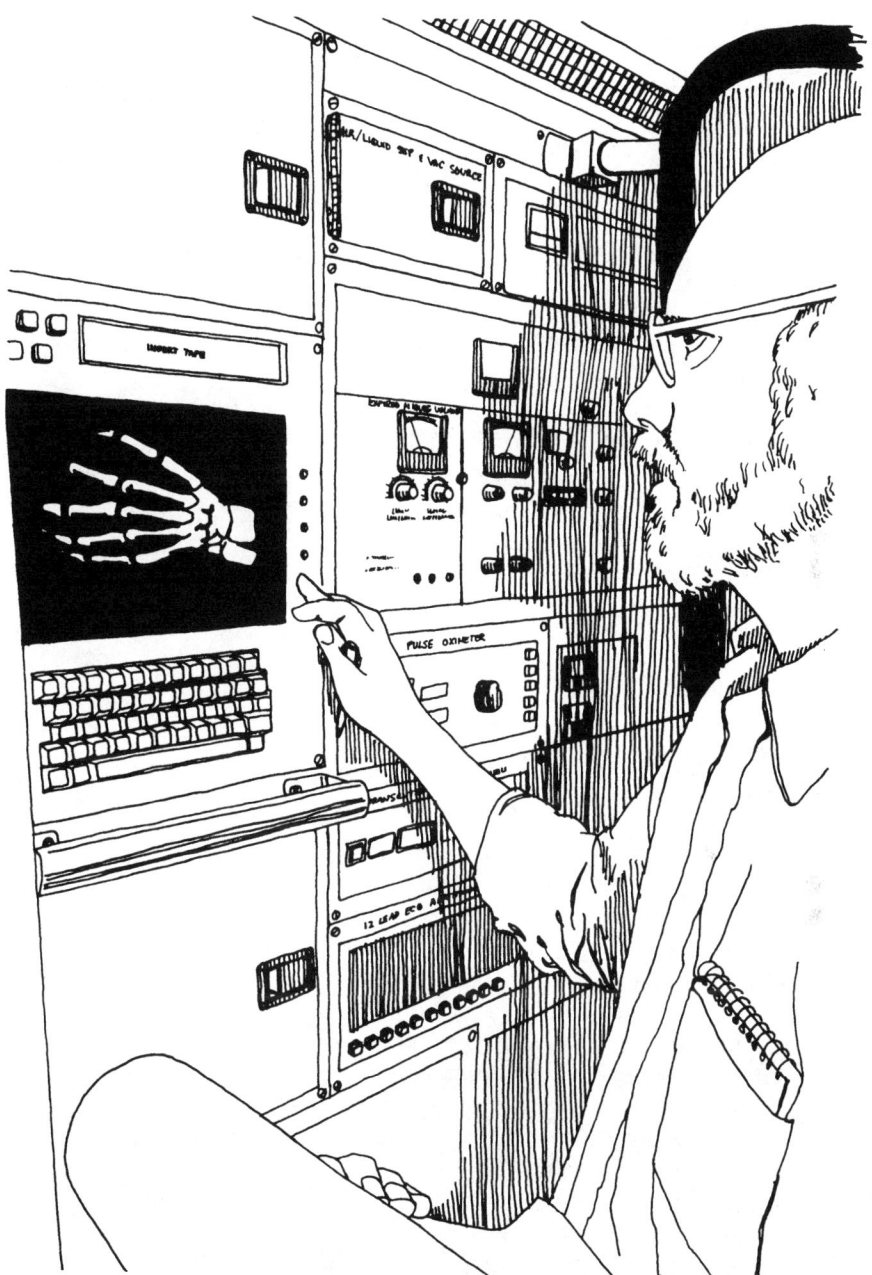

The medical officer on board the space station conducts a lot of biological research with the goal of solving many of the perplexing problems that face deep-space voyagers, such as bone loss and the loss of red blood cell mass.

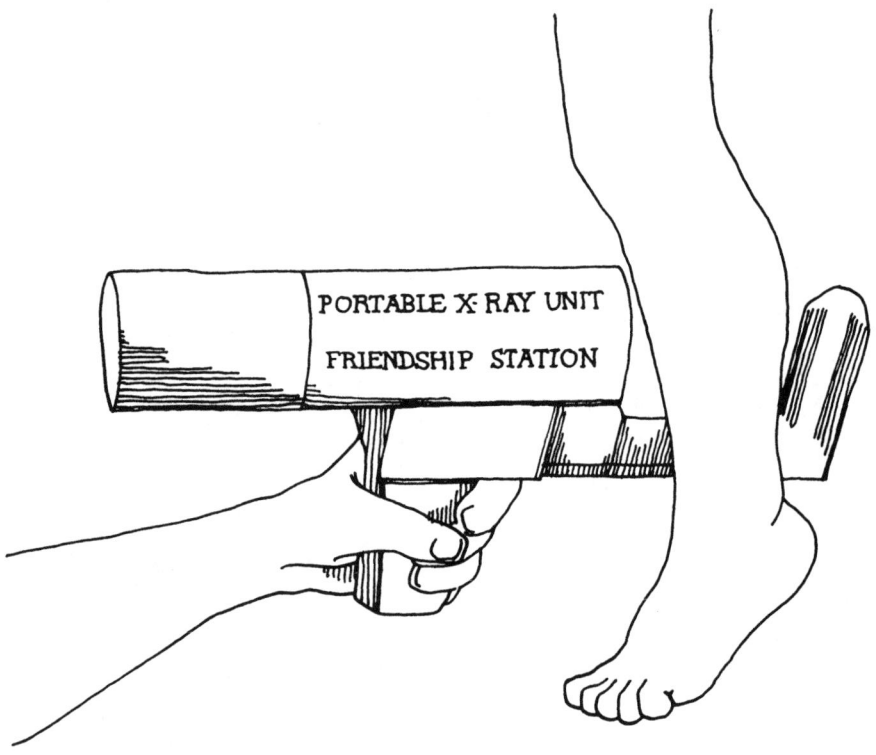

The portable X-ray used on *Friendship* has applications on Earth in dentistry, in orthopedic surgery, and in airport security checks.

tionship were all being monitored. It had been determined that exercise during space flight was an absolute necessity.

In neurophysiology, Beth's team was measuring touch, smell, hearing, taste, sight, nerve transmissions, sleep, and motion. To be effective, astronauts must keep in touch with the world around them.

Finally, they were studying biology. They were looking at the body's rhythm of temperature, blood pressure, the internal clock, and manual dexterity. Also, they were studying the effect of zero-g on single human cells.

One of the diagnostic tools used aboard the space station was the bone stiffness analyzer developed by both NASA/Ames and Stanford University. It measured bone mass and stiffness and was used for maintaining bone strength during long missions when bones tend to atrophy. It could also be used in treating bone fractures and osteoporosis.

Another NASA development on board was the new portable X-ray, which was about the size of a Thermos bottle. This lixiscope low-intensity X-ray imaging scope was a battery-powered fluoroscope and used less than 1 percent of the radiation required by conventional X-ray devices.

"Well, do you kids all have a better understanding of what goes on here in life sciences and Animal House?" Beth asked.

"I know I do," Mary answered. "Especially since so much of your work ties directly into my thesis, I think I'm going to want to spend a great deal of time picking your brain while I'm up here, Dr. Tippett."

"It's Beth, Mary, and I've been looking forward to having you on board ever since I first heard about your studies of space stress. I have a lot of ideas I want to share with you. So after you get done with your orientation let's be sure to get back together."

"Thank you, Beth. I'll be here."

"Now," Beth said, turning toward an enclosure, "let's see how friendly Daisy can be today. Have any of you ever met a space monkey?"

5

The Garden_____

"Well, I'm sorry Daisy pouted so. Perhaps tomorrow she'll be friendlier. She has her bad days, too, you know." Beth had led them from Animal House into the garden module, where they now gathered around a giant pumpkin growing suspended in space. Tethers kept it from floating around the room.

"We're constantly amazed at how some of our plants respond to weightlessness. We're planning on making a pumpkin pie after we carve this guy up for Halloween.

"Don't worry about the masks; you'll get used to them after a while. We wear them because of the decreased oxygen level and because of our cloud fog system, developed by Tom Mee. We get better plants this way. So if you ever come in here to visit Brother Pumpkin, just be sure to snap on your breathing mask and keep the double-zippered airlock barrier closed. That way we won't let any of this good cloud fog system air out. Questions?"

"I have two questions," Wayne said. "First, is this really a practical way to supplement our diets up here? Also, what do you have growing?"

"First of all," Beth answered, "you have to understand that we're trying to decrease our dependency on the consumables that we have to bring up on the shuttle—things like food, water, and oxygen. Not too far down the road, we'll be making very long voyages, like out to the mining camps on the asteroids. We just can't afford the weight cost of taking everything along. Seeds don't weigh much, nor do embryos.

"On our trip out to Mars, the gardens the crew will have on board will be for supplemental food, but mostly they will be there for psychological reasons. The trip is only going to take a little over six months going and about the same coming home. The crew will enjoy the fresh radishes, onions, and all the rest, and especially the fresh flowers. But the really big benefits, and the breakeven point with this new controlled-environment agriculture, will come on the long voyages. Then all we'll have to do is supplement the on-board vegetables and fresh protein with condiments, seasonings, and things like B-12."

"One of the really neat things to think about," Billy said, "is that as we journey out into the stars we will be taking our living things from Earth along with us. We can farm the universe, so to speak.

"You know, I read an article once that talked about orbiting some large space mirrors over Mars and melting the Martian ice caps. The water vapor from the melting ice would trap incoming solar heat to raise the temperature. The water vapor molecules would also complement the peroxides in the Martian soil to release oxygen. They figure that would take about 400 years. And then we could turn the red planet green. We'd have a pretty good-sized vegetable patch up there."

"I remember," Mary said, "somebody telling us in training that you were in phase two of your research in this module. What does that mean?"

"Okay. Well, we have to build this technology one step at a time. And in phase one we were just trying to catch up with the Russians. We had to experiment with plant root growth, establish our water-cycle requirements, and decide on our waste-system chemicals. That was the basic purpose of phase one.

"You know, cities sprang up when mankind first stopped being nomads and discovered that they could beat the seasons and starvation by planting seeds in one spot. The result was the development of agriculture, which was a key in the development of civilization itself.

"We are now entering a similar but different form of that equation. Here we are on the move once more, but now we intend to take our goodies with us. But I don't want to bore you with all this."

"No, go on, this is interesting," Mary said with a sparkle in her eyes.

"Okay, well, that first phase lasted four years. We're a year into phase two and we've scaled up the results of phase one far enough to sustain two of the crew. We're looking at plant nutrient pathways and carbon, oxygen, and water transfer rates into and out of the system.

"We expect to be free of phase two in three to four more years. Then we'll take the big plunge and go into an operational test on a prototype system to support two people entirely to verify its livability. They will be completely supported by the CELSS."

"'Scuse me?" Mary asked.

"CELSS, the Closed Environment Life Support System."

"Oh."

"You've undoubtedly noticed that this is a split-level module. We have a basement—or an attic, anyway you want to look at it. Some of our plants like artificial light, but others demand normal sunlight. An old farmer's trick we're experimenting with is spraying sugar water on sample flats. Your great-grandmother often used this 'artificial sunlight' to start her early spring seedlings. Downstairs we have our greenhouse with quite a number of windows to let the light in. Come over to this opening in the deck and you can see what I mean."

"Hi, Dr. T."

"Oh, hi, Dave. Mind if I show off your domain to our new kids?"

"No, be my guest, come on down."

One by one the group pushed off down to the low room below. Beth showed them the reflectors outside on outriggers that angle the maximum amount of sunlight in. Since *Friendship* goes through a day/night cycle about every 90 minutes, the reflectors are controlled by HAL to squeeze every last drop of liquid sunlight out of each orbit. To supplement the natural sunlight there are automatic hothouse Gro-lites that come on when they are on the backside of Earth. After a "normal" day's worth of sunlight, the automatic shutters are put over the greenhouse openings and the reflectors go into an energy-storage mode to gather solar rays for the next day's darkside passes.

"What are those lovely lavender flowers over there?" Mary asked.

"Those are water hyacinths," Beth answered. "They're my favorite friends here in the garden, along with my grandmother's hybrid tea rose that I brought up for company."

"Are the water hyacinths just for morale?"

"Oh, no. They're an experiment, too. They show great promise of becoming an important part of our CELSS system. For several years Billy Woverton and Becky McDonald at NSTL in Bay St. Louis, Mississippi, experimented with them to treat wastewater, and we're now carrying that one step further. Water hyacinths are great for purifying domestic and industrial wastewaters down below, and they can be ground into fertilizers, too.

"Other plants that are part of our CELSS project here are sugar beets, lettuce, snap beans, wheat, soybeans, white potatoes, and sweet potatoes.

You know people around the world eat about 7,000 different kinds of plants for food. There are about 250,000 plant species in all. We really only depend upon 15 of those. Strangely enough there are at least 75,000 plant species on Earth that are edible, and many of them are probably superior to the crop plants we now depend on.

"Nutritionists are looking at four different groups of space-farm plants: protein, fat, carbohydrate, and vitamin accumulators. We'll get the vitamins from parsley, dill, spinach, radishes, tomatoes, cabbage, peppers, onions, and lettuce. Protein will come from the soybeans, as well as peas, beans, rice, wheat, and peanuts. The sweet potatoes here will give us the carbohydrates, and so will the wheat, sugar beets, carrots, cabbage, rutabagas, and kohlrabi. The Russians grew most of these in their space stations of the late seventies and the eighties.

"An interesting area of research here in the garden is in tree growth," Beth said. "Without gravity to help develop their root systems, we've had less than spectacular results with trees. What we have developed, however, are some of the most beautiful bonsai trees in existence. You can see by our bonsai chamber over there that it's very popular with the crew. You should see the one that Mitsue and Shoji have in the Japanese module. It's really a grandfather tree."

Suddenly the scream of a siren filled the air. The newcomers turned cold with fear, especially when they saw the worried looks on the faces of both Beth and Dave. The CRT at the work station flashed rapidly: "HOLE IN CABIN, HOLE IN CABIN!"

HAL's synthesized voice, soothing yet authoritative, came immediately over the speakers. "We have a hole in cabin emergency in Mod Seven. Please do not become alarmed. Don your emergency oxygen masks at once and secure all hatches in your modules. Repeat, don your emergency oxygen and close all hatches. This is a hole in cabin emergency. We have a penetration level-three hole in Module Seven. Remain calm and stay in your module. Our safety and secure squad is on the scene. One moment, please."

Beth tore open the emergency oxygen locker marked in red and gave masks to everyone. Her eyes traveled around the control board for the garden's systems to make sure that its integrity had not been compromised.

"Everybody be cool," she said quietly. "Stu and the Mod Squad will seal the hole as quickly as they can so we can all get back to normal. If you want to take a look at the scene of the action, Module Seven is over there through that porthole. See the one with the observation room on top of its outboard node? That's to help the crews who maneuver the remote manipulators.

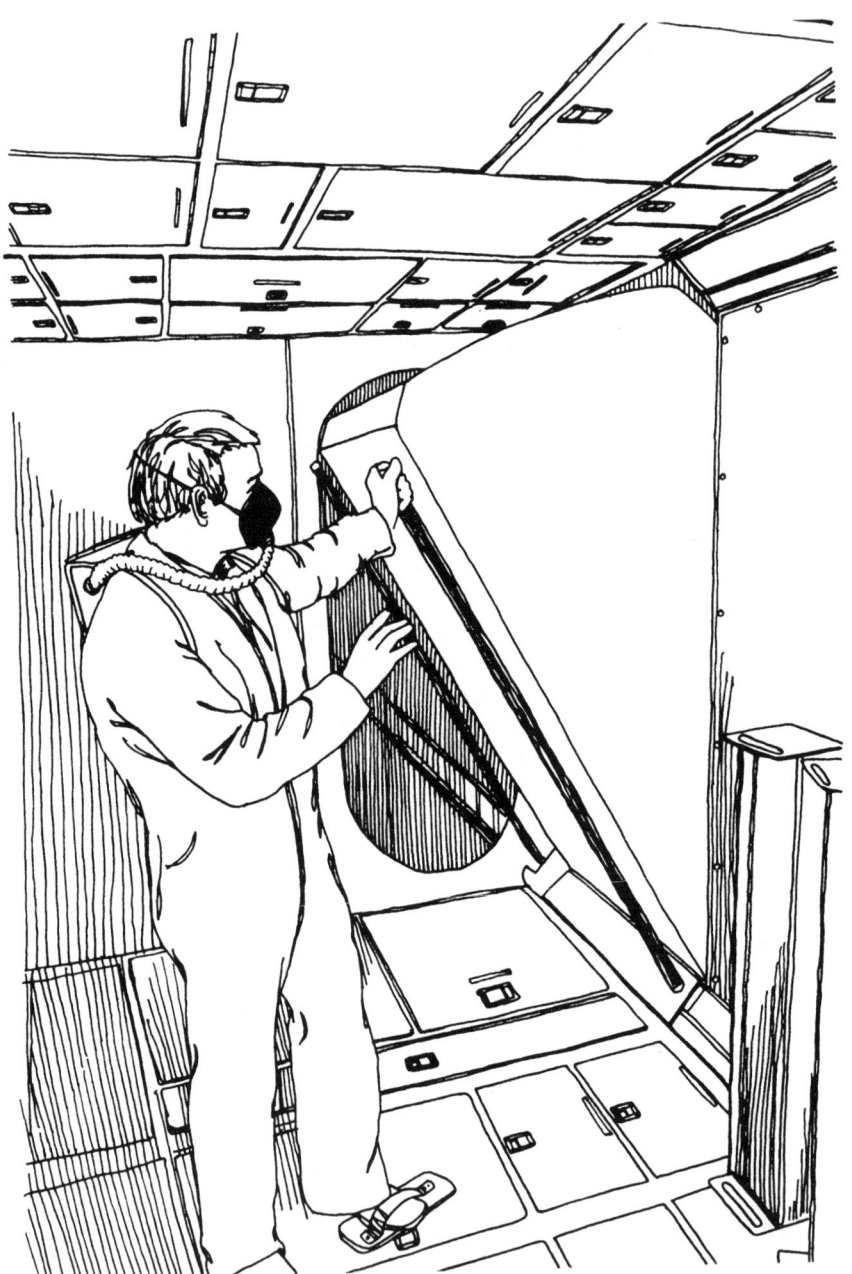

An emergency-team crew member removes a rack to get to the micrometeoroid leaker behind it. This is a critical design feature of the space station modules. Note that foot restraints are used as an aid to moving a mass this large.

"Somebody had the smart idea to set up some of the RMS stations inside so that not all of the work had to involve EVAs. Unfortunately we don't have the micrometeoroid shield quite right yet, and it sticks up like a fat man on a you-know-what, as my Irish grandfather used to say. We'll figure it out sooner or later. These alarms get us all on edge and we certainly don't need that."

None of the rookies had said a word. They were all terribly frightened. They had read of space disasters from grade school on, and *Apollo 1, Challenger,* and some of the Russian disasters suddenly seemed very real. But they knew that the safety of the station had been impeccable up to this moment, and it had never been seriously crippled for any length of time.

"Attention, please," HAL said smoothly. "The hole in cabin emergency is clear. Repeat, the hole in cabin emergency is clear. Oxygen dump is complete in Mod Seven, and the repair crew is sealing the module from the inside. Other modules may now resume normal activity. You may resume."

Everyone breathed a sigh of relief; there were a few nervous giggles.

"You never get used to it, folks," Beth said quietly. "It's just something that you have to learn to deal with if you are to serve out your tour. If today's emergency frightened you too much, then I really think you should consider riding back with Smokey and the gang when they leave. If you think today was scary, think of what it's like when we don't have a shuttle here to ride back in."

"Beth, are you still in the garden?" a worried voice on the intercom asked.

"Yes, we are, what's up?"

"Better come quick. Daisy seems to have gone into a coma."

"Damn! I'll be right there." Turning to the rookies she said, "Look, can you all find your way back to Mod One? I'll meet you there after I check in with Daisy. She did this once last week, too. We seem to be overloading her with something in our drug experiment. Grab a snack and relax for a while if you can."

And with that the three newcomers found themselves alone, with Dave now off again puttering in his garden in the basement. They looked at the hatchway through which Beth had left. For the first time they would try to maneuver through the space station on their own.

6

The Observers_____

Barney Caldwell tracked the rookies down as they made their way on their own through the space station. While Beth tried to get Daisy stabilized, she had asked Barney to spend some time with them. He was happy to have the opportunity to tell the rookies about his work.

Working under a grant from the National Science Foundation Barney and his group had chosen to study a tiny mystery planet, which they called Hydor, in a rather small galaxy. Among other things they had discovered that it had a very large companion moon in relation to its own size, that it tilted at 23.5 degrees, and that it was in an undistinguished group of planets that circled a rather ordinary small star.

They also knew that its diameter was about 12,756 kilometers and that its magnetic poles wandered. Their paleomagnetists were studying the possibility that its magnetic field even reversed itself from time to time.

Billy asked why they had chosen to study Hydor.

"We had 100 billion galaxies to choose from, with maybe 10 billion trillion planets," Barney answered. "Someone once estimated that if only one star in 100 is single like our sun, and if only one out of 100 such single stars has a system of planets like ours, and if only one out of 100 of these has an Earthlike planet, and if only one in 100 Earthlike planets is the right distance from its star, and if only one in 100 of these has the right chemistry of water, land, and air somewhat like ours, then there are 10 billion planets out there suitable for life as we know it. We are attempting to define exactly what that proper chemistry and mixture is and how it all works together. Someday people on Earth may need to escape to another planet, and that is our primary motivation. So we zeroed in on Hydor.

"Hydor is surrounded by belts of radiation that probably result from the plasma particles given off by its star. You know, a star isn't a solid body like a planet; it's only a ball of gases held together by its own gravity. Some of its layers even revolve at different speeds. Hydor's star is 69 percent hydrogen and 28 percent helium. The remainder is oxygen, sulfur, carbon, nitrogen, iron, magnesium, and silicon. As a matter of fact, a square mile of this star's light would only weigh three pounds."

Wayne interrupted, "How big is the star in relation to Hydor?"

"Let me try to remember," Barney said. "Yes, I believe our latest estimate is that it's about 332,000 times larger than Hydor."

"Holy Moly!"

"Yes, though it's a minor star, next to its small planet it's a giant. I think maybe you can begin to imagine how tightly it holds the small planet in its grip. Hydor even lies within the outer atmosphere of the star, what we would call its corona; some people also call it the solar wind. Someone once said it's like trying to push a robin's egg with a bulldozer. That's how tenuously Hydor hangs on to life as it runs around just an arm's length from disaster with its host star.

"The planet itself has a solid core that we are pretty sure is either nickel or iron, and that core appears to be floating in a molten liquid of some sort. Only about 3 percent of Hydor's surface is suitable for what we call agriculture; 24 percent is mountains and deserts, and another 2 percent seems to be made up of trees. Matter of fact, Hydor's mountains and canyons are rather puny compared to what we have nearby on our own Mars.

"Its skies are nitrogen, which gives it a rather distinctive hue because of the way the starlight is scattered by the molecules in its atmosphere. Incidentally, in relation to Hydor the atmosphere is only about as thick as the skin of an apple. So it's a very fragile planet, really an awfully lot like that robin's egg, rather than the rock it appears to be.

"Hydor only has five million billion tons of air as we know it. About 72 percent of its surface is composed of a hydrogen-oxygen compound that turns to a solid below zero degrees Celsius. It is a liquid up to somewhere over 100 degrees Celsius, when it turns into a vapor or a gas. We figure that it has 326 million cubic miles of the stuff. Only 3 percent of that is fresh and about three-fourths of the fresh is frozen.

"It has the usual weather systems that we see on our computer models for planets in this position relative to their mother stars. About 9 million tons of rain fall on it every minute, and some 1,800 thunderstorms rumble across its surface at any given moment. It's a dynamic little planet during its storms. Our best guess is that about 100 lightning flashes occur around the planet at any given second. They are more than twice as hot as the surface of its star. And it would take a cloud ten to twelve days to circle Hydor if it didn't dissipate first.

"We figure nearly 4 million tons of star dust fall on the planet every year, much of it as old as its own solar system, which is about 4.6 billion years old. But since Hydor weighs 6.6 billion trillion tons, it hardly notices that it's putting on weight."

"Are there any signs of life on Hydor?" Mary asked.

"Oh, yes. As nearly as our most sophisticated sensors can figure out, there are 30 million species of plants, animals, and microorganisms on the planet."

"Wow!"

"Yes, it's a prolific little island. They seem to enjoy procreating, even the plants and microbes. And they all seem to be in constant conflict with one another. There are those that eat and those that are eaten. Everything is consumed and changed by something else. And everything is in constant motion even within the interior of what we would call Hydor's rocks."

"Wait a minute," Wayne broke in. "How in heck can you tell that from your orbiting sensors above Hydor?"

"If you'll look out that observation window over your shoulder, you just might be able to see it right now," Barney answered. "That planet you're now skimming around, with its nitrogen blue skies and surface covered with that hydrogen-oxygen compound that turns solid, and that you primitives might call water and ice, is the one that my team and I are studying from our lofty perch. The ancients might have called it Hydor—the Greek word for water."

Everyone enjoyed a warm laugh together as they looked out at the blue, green, brown, and white "Hydor" slowly rotating below them. They were seeing Earth through new eyes.

"But to get back to business, students, Beth and Stu asked me to introduce you to our earth sciences department so you could sort of know your way around our module when the time is right. Since most of our equipment is located off the station proper, I can do a pretty good job of filling you in right here where we can all be comfortable.

"What say we do a midnight requisition job on those fresh donuts you brought up in the logistics module? I think I saw someone hide 'em in this bin. Why don't one of you fix us some coffee? It'd be good practice for you."

Later, the group lounged comfortably in an unusual cluster of shapes and attitudes as the rookies began to relax and let their bodies seek their own positions.

"As I started to say," Barney began, "much of our equipment in earth sciences is not on board. In earth sciences we actually have three basic sets of eyes trying to gather as much data as we possibly can. More importantly, we try to keep it all synchronized so that we can get a long and a short four-dimensional view of what's going on—the fourth dimension being time. You know, I should say we have four sets of eyes since we also coordinate all of our data with observations being taken on the ground at the same time.

"Eye number one is on our polar platform. We have a synergistic grouping of instruments that are coordinated with the other measurements taken with our other eyes to give us a good view of Earth as a total system. Vandenberg services our polar station, the *Richard E. Byrd,* named after the first person to fly over both the North and South poles. Our polar platform is just about the same size as Byrd's first plane, the *Josephine Ford.*

"Our *Byrd* in space is in a low Earth orbit altitude of 380 nautical miles and at an inclination of 98.2 degrees. It circles the Earth every 100 minutes or so. This orbit has a sixteen-day repeat cycle and is sun-synchronous with a 2:00 P.M. nodal crossing to avoid the sun glint off the water. By going over the poles the *Byrd* can look at the entire world rather than just one swath of it as the space station does, and as most of our other spaceships have always done. As you know, the station is on a 28.5-degree orbit above and below the equator, so we really go over only the tropics. The highest inclination we can launch toward at the cape is 57 degrees so that we do not launch over populated areas.

"Our polar station is unmanned but is serviced by the shuttle out of Vandenberg in conjunction with one of our orbital maneuvering vehicles. We usually command the OMV to hook on to the platform and bring it down to 150 miles or so. The shuttle crew and payload specialists can then recalibrate instruments, refurbish and repair pieces of the *Byrd,* and add new sensors or experiments as well as unhook the ones that they want to bring back down Earthside. Then the OMV tug simply pushes the *Byrd* back up to its parking place at 380. Whenever possible, though, we simply leave the *Byrd* where it is and take whatever we need to take up on our "smart machine," the OMV. We have the capability to replenish some of the consumables and to do basic servicing that way. Am I covering too much ground for you too quickly?"

"Not at all," Wayne answered, "this is all very fascinating. What instruments does the *Byrd* have on board right now?"

"Rather than dazzle you with exotic names, let me just say first that what we're trying to do is get a handle on how Earth works. For example, twenty years ago we could forecast weather a week in advance. Later we pushed that out to two weeks. But we needed not only to go for longer ranges on our projections but also to understand the whole system in more finite detail.

"We're trying to grasp how the clouds, winds, moisture, evaporation, tides, currents, and land systems all work together. We need to 'see' how it all interacts, and we need to understand the Earth's global electrical circuit. Now we just see chunks of it at any one time. And an awful lot is at stake—not just knowledge itself, but with our exploding population it could even buy us some time before we have to evacuate this small planet. Hurricanes, storms, droughts, tornadoes, and floods are daily occurrences around the world. So are famine, crop infestations, and loss of habitat for our wild plants and animals.

"We animals who have learned to fly spaceships must realize that every single action in the dynamics of the global system of life has an impact on us. You know, even the lowly rosy periwinkle has become precious to us in fighting leukemia. So you can see the truth of what the Nature Conservancy says, 'If we save wild species, maybe they'll save us.'

"We simply have to start respecting the sanctity of plant life as well as animal life. Without plants and trees we would be in serious trouble. We need the oxygen they give off for our very cells to work and to stay alive. Half of all the rocks, minerals, and even the Earth's crust is oxygen.

"We've been lucky so far. Thanks to the photosynthesis of the plants and trees, as well as the decomposition of water vapor in the upper atmosphere caused by the ultraviolet rays from the sun, we've been able to maintain roughly the same amount of oxygen in this oxygen tent we live under called the sky.

"Sorry to go on about this, but it's so vital to our survival. Just this year alone, a forest the size of the state of Virginia will disappear forever from the face of the Earth. Forests are disappearing at fifty acres a minute. More than 40 percent of the world's rain forests have already been cut down or burned down. And studies show that this is already reducing rainfall and increasing temperatures, probably even affecting our global weather patterns. Tropical forests cover less than 10 percent of the surface of the Earth, yet contain nearly half of its plant and animal species.

"Incidentally, a full one-third of all of the plant life on Earth is found in the Amazon basin. We in the United States get a great deal of the oxygen we breathe from the Amazon area. Thank goodness some countries like Brazil are doing something about the destruction of their rain forests. They've passed a law that no more than 50 percent of their Amazon forest can be destroyed. Even that is too high a price to pay for progress, but at least it's a start.

"The tropical forests also contain the largest pool of carbon on the planet. So by destroying them, not only do we take away the very critical oxygen factories they represent, but by burning the trees and releasing their stored carbon into the atmosphere along with the other carbon-based fuels we already are throwing up there every minute of the day, we further rot the thin skin of our apple.

"But it's not just the tropical rain forests that are in trouble. Even the Swiss have a problem, what they call the *Waldsterben,* a German term for the dying forests. The forests in Switzerland are dying from nitric oxide emitted by automobiles, from hydrocarbons given off by factories, and from 'acid rain.' Over one-fifth of the Swiss forests already suffer from pollution damage. And with nearly 300 million trees in the country you can see what a huge problem it is.

"So, anything that happens on Earth, physically, or that happens to Earth, physically, from outside influences, that's in our ballpark.

"We have to learn to predict, not merely describe, the solid body geophysics, the land, the fluid portions of the environment, and the sun. We do that by studying the space plasma physics to try to understand the near-space environment and the whole solar-terrestrial marriage with its ultraviolet, visible, and X-ray bands of input into the system.

"We measure energy input, and we're defining the storage regions as well as the discharge of this energy toward Earth. Some of this is done with tethered subsatellites. See that thing that looks like a fish line disappearing down below *Friendship?* It's made of Kevlar and extends down below us for more than 175 miles.

"Another important area we're looking at is monitoring the atmosphere. That was first proposed at a United Nations conference way back in the eighties, even before *Friendship* was off the drawing boards. They discovered that unless air pollution was curbed there would be a ten-degree temperature rise around the world that would melt the polar ice caps and cause worldwide flooding. One expert even said this 'greenhouse effect' could bring about the extinction of mankind.

"We're studying the various layers of the atmosphere just like it was a wedding cake. We're looking down through the exosphere, the ionosphere, the mesosphere, the stratosphere, and finally the troposphere.

"The troposphere contains almost all of our weather and goes from the surface of the Earth up about ten miles high at the equator. About 75 percent of our atmosphere's in that layer. Our LIDAR radar system measures the water vapor there. Where the troposphere leaves off the stratosphere takes over—it's about twenty miles thick—and so on up the hill.

"Our sensors and data system also support research in atmospheric dynamics and radiation. The atmosphere, the land, and the ocean all interact with one another. Temperatures, winds, and moisture all play a part in the entire system of global weather. For example, a sea surface temperature change in the subtropical Pacific can later have a very substantial effect on the winter weather in the eastern United States.

"So we need to study the oceans, their temperatures, circulation patterns, and how this all influences the food in the sea. We also need to study el Nino. It occurs every three to eight years and can have a dramatic effect on the weather and the food chains. We use a free flyer called TOPEX for much of this research, and we do a lot of it right here on *Friendship* as well as on the *Byrd.*"

Barney went on to say that two more polar platforms were being planned to handle the demand for research from space. Private industry had begun to use space-based sensors for oil exploration, fisheries, and other applications.

The LIDAR sensors can measure cloud-top heights, cloud-top winds, aerosol winds, surface and cloud-top pressures, tropospheric temperature, and surface albedo. Profiles can be made of water vapor, clouds and aerosols, stratospheric aerosol and ozone, and cirrus ice-water distribution.

"Microwave equipment with its large antenna," Barney continued, "does a lot of both active and passive measurements. It looks at surface wave imagery, surface wind speeds and direction, wave heights, warm core shedding, and sea ice mapping. It monitors sea surface temperatures, surface wind speeds over the oceans, atmospheric temperatures and humidity, polar ice boundaries as well as sea ice extent, including the percentage of open water, first-year ice, and multi-year ice. It also measures and maps precipitation over the oceans.

"We're trying to see the three dimensions both vertically and horizontally at the same time over the entire surface of the Earth. Like the man said, every day is a new day, and so we have our hands full accumulating new knowledge. Matter of fact our data system operates on 300 megabits—that's 300 million pieces of information."

"What about Earth itself, I mean the land mass?" Billy asked.

"Yes. Well, of course we measure things like vegetation just like we did with the old Landsats in the early 1970s. Those gave us our first real insights into the land dynamics. The energy balance of the planet is determined in no small way by the albedo, which is the reflected light and radiation from the surface of the globe. That changes with the amount and type of vegetation. We're studying this all year long as the seasons change and the atmosphere changes patterns. That way we can pretty well forecast global agriculture production and timber harvests and how to use range land for feeding our cattle.

"We're also monitoring long-range urbanization, industrialization, deforestation, and ranch land overgrazing, trying to get a handle on the whole question of waste management and the creation of uninhabitable areas such as deserts.

"Our hydrologic cycle research watches lakes, rivers, reservoirs, snow cover, ice, and aquifers for research management purposes. Soil moisture, evaporation, precipitation, erosion, and runoff all affect agricultural production as well as the energy balance. Evaporation also affects both weather and climate. We can do some amazing things nowadays with water retention and release when necessary.

"Now to the wet rock itself. You all know that the crustal dynamics are very important. I think the recent tragedy in California during the triple quake reminds us all how fragile our toehold on this small blue ball really is. We are studying the stresses within the Earth as well as the interior rheology and subsurface structures. We're monitoring tectonic plate motions with the help of our lasers and trying to predict and detect earthquakes before they happen. Our visible and infrared sensors are helping us map the surface structure.

"Our geosciences packages not only help give us a better picture of global rock distributions and geologic structures but also help detect and monitor volcanism, landslides, and the crustal dynamics I mentioned. We also watch other events with global and continental implications, such as insects and diseases, forest and range fires, oil spills in the oceans, droughts, floods, and on and on."

"Good Lord," Billy said, "it sounds to me that if it happens on Earth, someone or something up here is watching it and measuring it."

"That's about it. And I haven't even mentioned the other eyes we have on our GEO platform. You must remember that our eyes on board *Friendship*, as well as the *Byrd* polar platform, are always moving at their orbital speed around the planet—west to east with the space station and north to south with the polar platforms. But with our GEO platforms 22,300 miles out, we're always poised over the same spots on Earth. That's why we have three of them up here. At any moment we want to, we can observe what's going on below us.

"The geosynchronous speed of our platform out there is 6,876 miles an hour, but that keeps us right on the spot we want to stay, or within a two-degree neighborhood. And we can shift our location when we want to, also. Incidentally, our GEO platform is one-tenth the distance to the moon."

Barney explained that GEO 1 is the *Arthur C. Clarke*, named after the British science fiction writer who first came up with the idea in 1947 of putting satellites in a geosynchronous orbit. GEO 2 is called the *Konstantin Tsiolkovsky*, named by the joint NASA, ESA, USSR, Canada, and

The *Arthur C. Clarke* geosynchronous platform travels 6,876 miles per hour at an altitude a bit over 22,000 miles and remains over the same spot on Earth at all times for constant Earth observations. Clarke, a well-known science fiction writer, proposed the idea in 1945; he is known as the father of the communications satellite.

Japanese team to honor the Russian who first suggested artificial earth satellites, and GEO 3 is the *Robert H. Goddard,* named after the U.S. rocket pioneer.

Friendship's goal was simultaneous global coverage in order to see things as they begin to develop. With weather, for example, the goal was to make monthly hurricane predictions accurate within 24 hours and 100 miles.

The 40-meter microwave antenna in the 118 gigahertz region was measuring temperature profiles, and in the 183 gigahertz area *Friendship* was working on water vapor profiles. They were beginning to be able to track small precipitating cells over water and underneath clouds in order to measure winds. The antenna was built at *Friendship* and then taken up to GEO with the OTV *Homer.* They were working on the next generation so they could map precipitation in convective areas. By building time histories of weather cell growth, they hoped to predict precipitation better.

Barney continued, "Thanks to a British physicist by the name of C. T. R. Wilson, who first came up with the idea, we now know that our giant motor that is Earth is driven by the thunderstorms that help maintain its balance of power. Negative charges flow from under thunderclouds to the ground below, and positive charges flow upward above the thunderclouds toward the travel lanes in the ionosphere. There the charge is very quickly distributed around the planet and flows back down to Earth through the fair weather windows in the atmosphere.

"We're working right now to answer the question of whether the thunderstorms serve as current generators or as voltage generators in the global electrical circuit—and what part, if any, these weather systems have in how the atoms in the cells of our bodies work. Scores of researchers around the world are studying these environmental influences on the body. Remember that the brain is also run by electricity. True, the body's electricity is not from the boiling thunderheads above us . . . or is it?" The rookies smiled, but their rapt attention was unbroken. "The body's electricity comes from a compound that our body produces called ATP.

"Scientists now know that the spin of Earth on its axis and around the sun, as well as the gravitational forces of the moon and the sun, have produced biological rhythms within us that affect our bodies and our behavior. The most obvious example of this is the jet lag we all feel when we make flights covering more than one time zone of the sun.

"Another important part of the puzzle that we are putting together at GEO is our ozone work. Ozone, as you know, is located above our cloud systems in the lower stratosphere and helps shield Earth from solar ultraviolet radiation. In low-pressure areas at those altitudes, the bottom falls out of the stratosphere, so to speak, and the ozone-rich air fills in the area.

By studying this phenomenon we can detect changes in the jet stream that usually trigger severe storms, such as tornadoes, and that have such a major impact on the weather systems as they sweep and swirl around the Earth.

"We're studying the areas where the ozone folds back over itself, causing ozone to be lost in the lower atmosphere, where it is eventually destroyed. The people at the universities of Iowa and Wyoming are our ground controls. NASA and Langley too. You may have heard of the work that Gerald Keating and his team did many years ago at Langley on this. They were the real pioneers. Knowing about the ozone not only helps us in predicting weather but it also saves the airlines a ton of money in fuel since they can route their flights to take fullest advantage of the jet stream's tailwinds.

"We're watching two holes in the ozone layers at the poles. The hole above the South Pole is about the size of the continental United States. We think the holes were caused by pollution and are letting in excess solar radiation that could cause a great deal more skin cancer and harm plant life."

"You seem to be keeping close tabs on the Earth from here on *Friendship,* as well as from the polar platform," Mary said. "Are you just duplicating that effort on the GEO platforms, or are you watching different things?"

"Fine question. Remember that what we are basically doing at our GEO location is looking down at the same location all the time and watching the Earth breathe and exhale below us and change its colors and its temperatures, much like a living person. At GEO we can study things over days and weeks rather than just the minute slices that our *Friendship* and the *Byrd* give us. And we can cover the entire Earth twenty-four hours a day with our three GEO watchtowers.

"We can measure the vegetation wilting in the hottest part of the afternoon and the effects that high temperatures have on crop yields during critical growing cycles for wheat and corn. These are some of the most important food crops in the world, so this kind of research is directly tied to the benefit of mankind.

"The water content of snow is another thing we measure from GEO with our microwave equipment. This can be very important in projecting springtime floods. Freeze damage and other thermal happenings can be tracked by the hour.

"As far as the oceans below us from GEO, we're studying the subtle shift of sea surface temperatures and ocean colors. Loop currents in the Gulf Stream in the summer can be watched thanks to color changes, and so can ocean fronts and eddies where fish tend to congregate. The natural food in the oceans quite obviously is where you'll find the most fish, which of course is important to know for food supplies around the world.

"Estuarial plumes shift within hours because of the tides and winds, and we need to keep an eye on those, too, to monitor sedimentation and pollution. We actually have our sensors set right now to take readings on all of these things every 30 minutes on a routine basis. Of course, when we want to we can command a 'constant on' mode.

"You know, all of this data that we are constantly dumping on Earth is staggering, and there are hundreds of scientists and graduate students, just like you people, working at universities all over the world every day to interpret it and to use it for a better quality of life for all of us."

"This is fascinating," Wayne said. "Next time I'm caught in a snowstorm I'll think of GEO and wonder why I didn't pay more attention to the weather channel before I started out."

"Yes, you really should. Our 30-day weather forecasts are now 95 percent accurate. One other GEO project that we play a part in," Barney continued, "is in the raw data from our satellites that are studying the space plasma. Nearly all of the matter in the universe exists as a plasma in one state or another. And our solar wind is an electrical plasma coming from the sun. It consists of electrically charged particles, atomic nuclei, and electrons in electromagnetic fields. The solar wind from our sun flows toward us at about 300 miles a second, and we've thrown a handful of electronic bottles out into this ocean to see what it is and to try to understand it better. Every day our lives are influenced by the solar wind. Weather, crop yields, radio and television reception are all affected.

"The primary storage areas for the plasma are in the tail of the magnetosphere and the electric currents that surround the Earth. The four satellites that we still have working and that have been out there in their worn orbits for quite some time now are in four different locations. They're all monitoring plasma composition, energy particle velocities, plasma waves, electric fields, and magnetic fields. A very fascinating bit of new knowledge."

"Are there any plans for a permanent manned station at GEO?" Wayne asked.

"Not right away, but we hope to have that capability in another five or six years. Right now it's just too damn risky, what with the solar flares and radiation levels. We plan on continuing the equinox flights that the press has been having fun with the past couple of years, but longer than that is out for right now. We've talked about taking up more substantial storm shelters for solar red alerts, but Congress is not too impressed with our payback schemes. They're too afraid of the bad press if we lose a crew at GEO from radiation. So for now, our twice-a-year visits are it."

"We've all read about them in the papers, but could you review the whole GEO flight thing with us?"

Barney outlined the purpose of the GEO flights. He spoke from experience. Earlier in his career he had made a flight up to the *Robert H. Goddard* to watch the weather over the United States.

The GEO platforms are in sunlight almost continuously except during the equinoxes each spring and fall, when they dip into the Earth's shadow for about one hour out of every twenty-four. Those are the only times they enter the Earth's shadow cone. And since the sun travels so quickly north and south at those times of the year, it causes violent storms because of the changes it triggers in the patterns of warm and cold air masses. When these equinoctial gales occur, it helps to have human eyes on board the platforms. The GEO platform visits are rotated among the three locations unless unusual readings come from the on-board instruments indicating that something unusual is about to happen.

"What's it like to travel up on the OTV?" Mary asked.

"Well, we first had to change out the fuel tank from the last flight by adding our new TRW propulsion module tank. They're brought up full on the shuttle so that we don't have to do too many tank-to-tank transfers up here close to *Friendship.* That's one advantage of using the 204 hydrazine. Plus it also has a long 'dwell time,' and we can store extra fuel for the OMV tugs. It gives us a nice clean burn for operations around the station.

"On my trip, Bobby Dalton and I were the crew. It's a six-hour trip, and we timed it so that we'd be arriving at the *Goddard* just after its dawn phase. That gave us plenty of time to dock before the one-hour night occurred.

"Now, let's remember that we're moving along at 18,000 miles an hour here on *Friendship,* and we want to rendezvous with a target moving at 6,876 miles an hour. In the process we want to change our altitude from 270 miles high to over 22,000 miles up.

"In order to make it into our transfer orbit we have to make a burn to pick up an additional 8,100 feet per second. Once we reach the high point in our new elliptical orbit we then have to do a second burn of 5,900 fps to turn the corner and circularize our new orbit. It's all a bit tricky, you know, and something that we do very, very slowly and deliberately. Once we docked with the line shack at the *Goddard*—"

"Excuse me, Colonel Caldwell," Mary interrupted, "why do you call it a line shack?"

"It's a term we borrowed from the Old West. A line shack is a haven that the old range riders used if a sudden snowstorm blew up and they couldn't make it back to camp. The shacks weren't fancy, but the cowboys always made sure they had dry firewood and some canned goods in them so they could wait out the storms.

"Up at GEO we have storms to wait out, too, only they're solar storms from the sun, along with cosmic radiation and charged particles. The name just sort of stuck.

"The shacks are really just modified versions of the old Spacelab modules that we flew back during the early days of the shuttle. There was no need to reinvent the wheel. They're 23 feet long and 13 feet in diameter so they could fit in the shuttle when they were first brought up here. Then we attached OTVs to them and transferred them on up to the high ground. That was back in 1998 and '99, if memory serves me correctly. Of course, we've got a special radiation shield envelope that deflects as much of the bad stuff as possible."

"You covered a lot of the work done at GEO a while ago, but what else goes on up there?" Billy asked.

"Remember that we can look at the sun 24 hours a day most of the year. That causes some problems for us in trying to get any decent sleep, but we have special shades we can use while we're camped out up there. The advantage of that is we can keep our solar telescope going most of the time. And speaking of that, I don't know if you're keeping up with our new *Wolf Pack* spacecraft or not."

"Not really," Billy answered, "I mean, I've heard of them in my class work, but that's about all."

"Okay. Well, at any rate, there are four of them now out on station. They're the same distance from the sun as we are here in Earthspace—93 million miles, give or take a bit. And they're separated at 90 degrees around the sun so that we have a constant eyeball on the solar flares and other solar features. And in stereo, no less. Not only do they give us an early warning system on the flares for when we're working in deep space, but they also provide us with high-resolution data about the poles of the sun. Since it only takes eight and one-half minutes for light from the sun to reach Earth, we're a great deal more under the influence of the sun for our entire terrestrial environment than most of us ever stop to think about.

"I could go on and on about our particle accelerators, coherent scatter radars, and plasma instruments that we have on board at GEO, but it's all pretty technical stuff if you're not into it. What we're trying to do, people, is to get a handle on this yellow star that lives in our front yard and has so much influence over virtually everything on Earth. And since the high-energy particles blown by the solar wind usually travel about 670,000 miles an hour, you can see why we're so anxious to keep close tabs on what's going on over there.

"Our sun chaser, *Little Boy,* is almost there now, and when he gets to within four radii from the sun we'll get a better handle on the solar density, velocity, magnetic field, and composition of the solar wind plasma. We're trying to learn where the wind whips up to the speeds we see here near Earth.

"*Little Boy* is also going to let us test out some of Dr. Einstein's general relativity notions. And then, in a nifty little maneuver the boys in the back room in Houston and at the Jet Propulsion Laboratory have all worked out, we're going to zip that sucker right out of sight by firing a thruster at its closest approach to the sun. The energy change will get it moving at a high velocity in no time at all. And we should be able to discover where the solar wind blends into the interstellar medium out in the great beyond.

"One other thing that I wanted to mention before we get off the *Goddard* platform at GEO. When the troops on board *Challenger* fixed the Solar Max satellite way back in 1984 they really gave us a great deal of new knowledge about our star, the sun. Max revealed a long-term decline in the solar constant; that's the sum of all of the solar radiation at various wavelengths that reaches the top of the Earth's atmosphere. One of our experiments gave us the rather startling information that if this downward trend continues it could trigger a new glacial age in about 150 years.

"Nobody was sure if this trend would continue or not, and that's another important thing that we're keeping an eye on at GEO. So far, it's too close to call. But if the present rate continues, the total solar irradiance would drop 2 percent in the next century, and the folks at JPL say that if it dropped by 6 percent the entire Earth would be covered by ice. Another fellow from the University of Maryland said that if it dropped by 13 percent we'd have global ice over a mile thick. That in itself is a pretty good reason to keep doing all of this research. And it really points up the sun's incredible influence on our small planet."

"Was it as complicated getting back down here to *Friendship* as it was going up . . . or over, whatever you call it?" Mary asked.

"Even more so. Don't forget, we were heading downhill into the fist of gravity. And we wanted to head back toward Earth and then hang a right, so to speak, so that we wouldn't reenter all the way into the atmosphere.

"We used aero-assisted trajectories so that the drag and lift could save us fuel as we brushed along the outer edges of the atmosphere. We did a burn to circularize our orbit and to raise our perigee to prevent going all the way back down to Earth. Our OTV's do have an offset center of gravity just like the old *Apollos*, though, to permit us to point the lift vector by rolling.

"It was one heck of an exciting rendezvous. We did it slowly and deliberately. It would have been awfully embarrassing to arrive where *Friendship* ain't."

Suddenly a voice crackled over the loudspeaker system. "*Friendship*, do you read us? Come in, this is Lunar Base Scott . . . need . . . your help!"

Barney rushed to the communications station. "Scott Base, *Friendship* reads you. Come in."

7

South Pole_____

"Scott Base, say again."

"Five bye, *Friendship*. We could . . . your help. How read me?"

"You're breaking up, Scott. Your S-band is wandering."

"Roger, *Friendship*. We're having trouble with our yaw axis. How now?"

"Better, Scott, much better. What can we do for you?"

"Say again, *Friendship*."

"Read you five bye, Scott. Is that you, Peter? How can we help?"

"Roger, *Friendship*. Yeah, this is Peter. Is that you, Barn?"

"Righto, Peter. What's up?"

"We've got a locking pin that's jammed on this damn thing, Barney. And we can't lock up with TDRRS for our comm link with Houston. Now that they've shut down Honeysuckle you seem to be the best game in town. Can you reconfigure and fill the link for us? If we can stay locked on you on your next pass back this way then we can satisfy that mission rule and still launch. Can do?"

"Roger, Scott Base. We'll file a waiver on our acquisition from *Wolf Pack* and turn our big ear in your direction. You're degrading pretty quick on this channel."

"Good deal, Barn. We'll talk to you on the other side."

Barney turned from the comm station to the waiting rookies.

"*Molly Brown* is due to launch into lunar orbit later this morning, and as you heard they've got a problem that we can give them a hand with. Let me talk to Houston on the brain machine and I'll fill you in later."

Entering a message on the keyboard, Barney filed a formal request with Houston to serve as the downlink with the lunar base crew. The message would dump automatically to the satellite and be relayed around the corner so that as the space station arrived over the United States later in the orbit they would have their formal approval. It was really just a formality, since all such decisions were left to the station itself, but red tape had not died down with the turn of the century.

Lunar explorers are always in constant voice and television contact with Earth by way of their S-band steerable antenna. It normally provides them with 174 degrees coverage in azimuth and 330 degrees in elevation. It features a 26-inch-diameter parabolic dish. But Murphy operates on the moon just as he does on Earth, and whatever can go wrong sometimes does. At these times, flexibility in the system helps. The space station had been used as a backup before, and since it was the crew's destination anyway once they left the lunar surface, parked their landing legs, and reacquired their aerobrake, this was really not too much of a deviation from SOP.

"There," Barney finally said, turning from the keyboard. "That's all taken care of. Now the kids will be cleared for takeoff. Imagine they're still number one on the runway up there."

"How long will it be before they get here?" Wayne asked.

"That all depends on how far the moon is from Earth at any time. On this trip it'll probably take them about 72 hours once they finally get reconfigured in the morning and head out our way. The trip's always shorter coming home because of the pull of our gravity."

"Reconfigured?"

"Yes, they drop their landing legs off the *Molly Brown* once they're in lunar orbit and then have to attach the 50-foot aero-assist skirt. The legs stay in a parking orbit for the next crew to use, and the aerobrake is really the key to being able to use this transfer vehicle system.

"Since the *Molly Brown* OTV tug is based here at the station, the aerobrake is what enables us to slow down in the friction of our marginal atmosphere at 200,000 feet and then roller coaster back up here to dock. It's a lot more efficient than using two vehicles like we did in Apollo where we had a command module to take us to the moon and then a separate lunar module to take us down to the surface and back up. Those were 'no-return' models and really expensive to use just once and then throw away."

"How much were the old lunar modules?"

"Oh, they cost about $30 million each. I think that included all of the test models and other support equipment that went along with them. But still they were really too expensive to be disposable."

"Does this kind of glitch happen very often?"

"No, as a matter of fact. The TDRSS III is very reliable, and this problem doesn't have a thing to do with TDRSS as such. It's just a hardware problem on the *Molly Brown*. Nothing we can't fix here in our friendly neighborhood service station once the crew gets here in a couple of days. Really just an inconvenience."

"Where are the TDRSS satellites now?"

"Well, one of them is southwest of Hawaii at 171 degrees longitude west, and the other is over the northeast corner of Brazil at 41 degrees west longitude. Our ground station is at White Sands, New Mexico. There are three 18-meter dish antennas there.

"With these two tracking stations we have almost continuous contact with the shuttles, whereas in the early days they could only talk with the Mercury and Gemini guys about 15 percent of the time. But our new state-of-the-art TDRSS relays are first-rate. We've only had them for a year. We get about ten years use out of these things and we really use them a lot."

"How long until they lift off the moon?" Mary asked.

"Not too long. Kathy will actually handle that end of it. Peter is the science officer on the crew and Kathy as commander does the flying. They have what you might call an equal opportunity marriage. They remind me a lot of Anna and Bill Fisher from the early shuttle days. You know, Anna was selected to be an astronaut first, and Bill had to grit his teeth and dig in and wait for another chance. Both of them were medical doctors, but Bill went back to school and got a degree in engineering so that he could up his chances of being selected. They both ended up flying.

"Kathy and Peter were in different astronaut trainee classes, too. Matter of fact, he was passed over three different times before he finally finished his doctorate in geology from Harvard. Then his stock really soared."

"How long have they been at Scott Base?"

"Just a week. Our last crew there stumbled on some good crevasses on the shaded side of Scott, but they were almost out of consumables and had to break off their work. Kathy and Peter took over where Andy and Phil left off. And they've had pretty good luck on their leg of Project Willow Stick. That name is somewhat of a misnomer since all of the water we're finding on the moon in these dark hiding places is really just dirty ice from passing meteors and out-gassing from the moon itself that has gathered there during the past couple of billion years. But it's a big find and has the guys on the ground excited.

The lunar module *Molly Brown* in Scott Crater at the south pole of the moon. The first U.S. husband-wife team is there prospecting for water for use in rocket propellants, as well as for hydrogen, carbon, and nitrogen.

"You know, we're going to use the water not only for drinking and sanitation but also for cooking up our cryogenic hydrogen and oxygen rocket propellant. That capability will really speed up our mining-base development program. We hope to get our mass driver operational sometime in the next ten years. But we can get into all that later; right now I'd better get Big Ear pointed the right way to support their liftoff."

Later, as the gangly station came around the western limb of the blue planet, Barney was able to establish radio contact with the crew at the moon's south pole.

"Good lock, Kathy. Houston has relayed your ascent PAD if you're ready to copy."

"Ready, Barn," the lunar commander replied.

"Okay. TIG 142034700 NOUN 76 55350000370 plus 0002, need A 47 in plus 37364 plus 05607 plus 58642 plus 56955, needle 465 is plus 00370, needle 546 is NA. Ignition 1 Rev late is 1440209, tug weight 10789. Over."

"Roger. Copy 142034700 55350000370 plus 0002 plus 37364 plus 05607 plus 58642 plus 56955 plus 00370, NA 1440209 10789."

"That's affirmative, Kathy. P32 CSI PAD follows. NOUN 11 143015060 NOUN 37 14438 all zips NOUN 81 0492 all zips. Need A 473 is 01818, 275 is 02780, AGS DELTA Vs plus 0492 all zips plus 0010. Over."

"Roger, Barn. 143015060 14438 all zeros 0492 all zeros 01818 02780 plus 0492 all zeros plus 0010."

"Okay, *Friendship*. I'm standing by for your up data link and the gyro torquing angles."

"Roger, *Molly Brown*. You can terminate your battery Bravo charge now and put your 02 tank 1 heaters and your hydrogen 2 heaters back to auto and dump your wastewater."

"Gotcha."

"You guys ready to end your lunar honeymoon, or is it a honeyearth up there?"

"It's a fascinating place to visit, Barn. But there sure isn't a lot of night life up here. We miss the old Golden Chariot. You been keeping up with the news from our old watering hole in Houston?"

"Not lately, gang. Guess we'll have to do it in person when this tour is over. Your gravity angle difference looks good now, Kathy, and there will be no uplinks to you this time."

"Very good, *Friendship*. Sounds like your patch link to head shed is working out all right."

"Other than the fact that they feel a bit left out, we're getting good data from them on you."

"Confirm Program 12 in, Barn."

"Roger. Your skirt ought to be just about overhead right now, gang. Have a good one."

"Thanks, *Friendship*. *Molly Brown* is on VOX. How do you read?"

"Loud and clear, Kathy. But your partner's awfully quiet today. Cat got his tongue?"

"I'm just drinking all this good stuff in, Barn. I don't expect to get back to Scott Base again and I'm just thinking how incredibly beautiful it is."

"Three minutes from liftoff, *Molly Brown*."

"Roger," Peter replied. "Checklist is complete, standing by for time-of-ignition minus two."

"Two minutes, troops."

"Roger . . . Mark, one minute," Kathy said. "Master arm is on."

"DSKY's blank," Peter replied. "Engine arm ascent. Twenty seconds."

"Looking good, Kathy."

"Good."

"Ten, nine, eight," Peter counted down. "Seven, six, five, arm."

"Go," Barney said quietly.

"Three, two, one, liftoff. And away we go."

"The engine has fired," Kathy confirmed.

"Looking good."

"Ignition guidance looks good down here, Kathy," Barney called.

"Thanks, Barnaby."

"Pitch over looking good," Peter said.

Kathy read out, "Three-ten feet above the lunar surface."

"Mark 30 seconds," Barney broke in.

"Fifteen-ninety-four feet."

"We're leaving town, *Friendship*," Kathy said with a laugh.

"And at one minute, yaw right 20, Kath."

"Okay, Petey."

"Two-sixty-four feet per second."

"Kind of wobbling around now," Peter radioed.

"*Molly Brown*, *Friendship*. Looking good at one minute."

"Heading right up the avenue," Kathy confirmed.

"Nice and smooth now," Peter added.

"Okay, *Molly Brown*," Barney laughed. "The lunar sky's all yours."

"Thank goodness. We forgot to look for air traffic."

"Don't you mean vacuum traffic?"

"Very funny."

"Six minutes. Verb 16, Noun 85. 57,000 feet." Peter called out, "Look at that crater down there."

"Beautiful, isn't it?"

"Five thousand feet per second velocity," Barney said.

"Looking good at seven."

"Engine arm off," Kathy instructed.

"All three data sources show Go," Barney radioed.

"Preliminary numbers show an orbit of 47 nautical miles by 9 nautical miles. Close to nominal, Barn," Kathy replied to *Friendship*.

"Good burn, kids. Go catch your skirt. Do you have your range rate readings?"

"Roger, Barn. Minus 331 feet per second."

"Good show."

The smooth-sailing *Molly Brown* space tug sped on as Barney turned back to the rookies.

"Now, where were we?"

"Tell us more about what Kathy and Peter were doing up there."

"Well, I mentioned the water search angle. That was an important square to be filled in our moon-base equation. Another was sampling the geology to confirm the preliminary data from our lunar mapping satellite that preceded this stage of our beachhead on the moon. That end of it was Peter's specialty.

"Somebody once said that it's a lot like the days when the transcontinental railroads were first opened in the United States. Refueling stops for water and wood were all part of the success of the system. They couldn't carry enough fuel and water with them to make steam and so they had to depend upon the local resources. That's basically what we need to do in space exploration, too.

"In the case of space travel, though, oxygen is one of the primary fuels. And we've decided that we can mine things, like ilmenite for its oxygen. It's a titanium ore, a black mineral. Actually, the lunar soil or raw dust is about 40 percent oxygen and we can use it for breathing, or process it for water, or blend liquid oxygen and liquid hydrogen together for making rocket fuel. Silicon makes up about 20 percent of the moon dirt, too, and we can turn that into glass, semiconductors, and solar collectors. Metals make up another 20 to 30 percent, and when we start separating them out we'll be able to use them for building some of the larger structures in space.

"You know, we're going to build large solar collectors from the materials we mine on the moon and beam the energy back down to Earth. It's a shame to waste all of this free solar energy all the time. Only half of one-billionth of the solar energy given off by the sun reaches Earth, and only about 35 percent of that is actually absorbed. The rest is reflected away.

"In just 20 minutes each day enough solar energy falls on the Earth to fill our energy needs for an entire year. All we have to do is smarten up and decide to collect it, store it, and transmit it down to the ground.

"It only takes 5 percent of the energy to lift up materials to Earth orbit from the moon that it takes from Earth. That's because the moon is near the top of our gravity well. So by mining the moon for the materials we need, we can conserve our own resources, and we can build space structures at a much smaller cost.

"One idea by Freeman Dyson, one of our old American physicists, is to build a shell around the sun itself, using material from Jupiter. That way we could collect 100 percent of the sun's energy for use on Earth and elsewhere in the solar system. They call it a Dyson Sphere. Seems a bit far-out to me, but so did this space station we're living on just 50 years ago, so who knows."

Billy interrupted. "Geology is one of my special interests. What else do we know about the moon?"

"Let's see," Barney reflected, "the typical Apollo sample gave us 14 percent aluminum, 4 percent iron, and 3 percent magnesium. There are also titanium, manganese, and chromium. A lot of the iron is a fine powder from meteor bombardments over eons of time. We can use our powdered-iron metallurgy after we recover it from the lunar soil. We plan on doing that on a conveyor belt with magnetic separators. There's a lot of glass in the lunar soil, too, and we can use that to make something like fiberglass on the moon.

"In Scott Crater Peter was looking for concentrations of hydrogen, carbon, and nitrogen. We think that they're concentrated in the polar craters where the sunlight doesn't reach. We'll know when he gets the samples back here to the lab. Our lunar mapping satellite, which I mentioned earlier, covered the entire globe of the moon and gave us some strong clues that we could find them there. Frozen volatiles at the lunar poles was one of its primary mission objectives during its year-long survey. It orbited every 118 minutes, and with its nadir-pointed science package it gathered tons of info. It dumped it to us down here at 32 kilobits per second and we caught it in our 34-meter antenna in our deep space network."

"Were all of the Apollo sites basically the same material?" Billy asked.

"Oh, no. There was quite a variation of material, which was to be expected since the planners tried to select places that were different. For example, *Apollo 11* landed in a relatively flat 'sea.' They found a lot of igneous rocks that are rare on Earth with elements such as free metallic iron and troilite. They also picked up a lot of crystalline rocks.

"*Apollo 12* landed just about the same distance from the midpoint of the moon to the west that *Apollo 11* had landed to the east. It made a pinpoint landing on the rim of a crater in the Ocean of Storms just 535 feet from the *Surveyor III* craft that had landed there two and a half years before. The rocks at this site were over 600 million years younger than those of *Apollo 11* and were predominantly crystalline.

"Of course *Apollo 13* never landed, because of the explosion on board that nearly killed the crew, but *Apollo 14* landed in what was *Apollo 13's* original target in the Fra Mauro formation south of the Sea of Rains. They found a lot of complex breccias rich in iridium, rhenium, gold, and nickel. One rock in their bag was 4.6 billion years old, although most were in the 3.9 billion range.

"The fellows in *Apollo 15* landed at about the same distance up the center line of the moon that *11* and *12* did east and west. It was targeted for Hadley Rille and the Appennine Mountains on the Marsh of Decay. It found the youngest moon rock of all the flights, one just 3.1 billion years old and another one 4.1 billion years old that they called the Genesis rock. Actually, if you look at the moon through a good telescope you'll be amazed at the height of the mountains that you can see even from here.

"*Apollo 16* landed on the western edge of the Descartes Mountains in the lunar highlands, part of what is the highest area of our side of the moon. Its rocks were 3.9 billion years old and about 75 percent of them were breccias. They also very unexpectedly found basalt rocks with a very high aluminum content. *Apollo 16* landed just to the south and west of the *Apollo 11* area.

"The last of our first moon flights, *Apollo 17*, landed in the Taurus-Littro Valley on the edge of the Sea of Serenity. It found 3.8-billion-year-old basalts that were 50 to 100 million years older than the basalts at the *Apollo 11* site not too far away. They found complex breccias there, too, and soil composed of orange glass spheres from volcanic fire fountains.

"One reason we were so interested in having Kathy and Peter follow up our search at Scott for ice deposits is that a month after the *Apollo 14* crew left the moon the suprathermal ion detector they'd left behind recorded 14 hours of clouds of water vapor that apparently was of lunar origin. Ions come from the out-gassing from volcanic or seismic activity, gases from a residual of the primordial atmosphere or from the evaporation of solar wind gases.

"Another tantalizing discovery was made by the *Apollo 15* and *16* command modules as they circled overhead. They collected data over about 20 percent of the surface of the moon, all within 30 degrees north and south of the equator. And they found very high levels of radon at Aristarchus Crater, Schroter's Valley, and Cobra Head. The Aristarchus readings were four times larger than the lunar average. Radon results from the decay of uranium and thorium. Thorium is used to produce uranium 233 for atomic fuels and is also used in magnesium. Once we get our permanent lunar bases up and running we intend to scout out those areas with field trips in our new moon expeditionary vehicles.

"We also intend to visit old Bailly Crater out west of our camp at Scott Crater. Bailly is the largest crater on our side of the moon. It's 183 miles in diameter. Its mountainous sides are 14,000 feet high; that's the same height as Mount Rainier near Seattle. And you could fit the states of West Virginia and Rhode Island inside it with room left over for Los Angeles, Chicago, Boston, Omaha, and Albuquerque.

"There are so many exciting places we want to search yet. Just on our side of the moon there are over 500,000 craters. Many of them are fresh and are made by the 100 or so meteors that hit the moon a year. We can tell this from our seismic equipment left by our Apollo crews. These hits range in weight from a little heavier than a tennis ball up to the size of a small car."

"What about the new moon spaceport?" Mary asked. "How soon will it be fully operational?"

"You can see it over there way out from that parabolic reflector. At least you can see part of it. The end of the tether is too far up to see the end of it; we're still experimenting with the correct length. Some of us feel it will end up being about 300 miles long. In reality we still think of this as the variable-g research facility. Once we get the artificial gravity thing all nailed down we're going to move it on out to its permanent orbit near the moon. And really only then will I be able to think of it as the moon spaceport. I guess I'm just too close to it right now.

"It'll be put into an elliptical orbit in order to transfer the momentum from approaching space tugs to those that are departing back here toward *Friendship*. As it dips close to the moon and then swings back out high above it, the use of the Kevlar tether for momentum transfer is going to save us tons of propellants.

"The moon spaceport will actually be the busy place out there then. All of the crews working down at the various outposts on the surface will transfer up to it before heading back here. We'll store hab modules, lunar rovers, cranes, and a whole pile of other equipment there. It'll serve as a staging area for all of our work on the moon. It will actually amount to another version of *Friendship* out there around the moon."

"Does it have a name yet?"

Barney chuckled. "Talk about a controversy. Nobody can agree on that one. It may just end up being *Moon 1* if they're not careful. Personally, I favor the *Apollo Station* in honor of all of the pioneer crews who hung their butts on the line to make the first trips out there. But I really don't expect to be asked."

"You mentioned outposts on the moon. How many are there going to be?" Wayne asked.

"No one really knows for sure yet. We're hoping that the samples Kathy and Peter bring back will signal the start of our fuel-processing plant at Scott Crater. If the samples prove to be as good as we think they will, then that will give us a big boost. We figure that we can have the pilot plant up and running in about three years' time. The basic technology has already been designed and tested in one 'g' so that all we really have to do is to transport the hardware up there and get it operational.

"After we get our fuel dump or service station going we'll be in good shape to use our raw materials out there to give us a jumping-off point into the rest of the solar system. I really can't think any farther out than that, but I suppose this might turn out to be the key for going throughout the galaxy. Kind of hard to think of such grandiose plans, though."

"What about other outposts?" Wayne asked.

"Well, there are quite a few on the drawing boards. There's the mass driver mining base, the lunar observatory, the far side radio astronomy facility, the core research base, the ag station, and on and on. What should I cover first?"

"How about the ag station? That's a new one to me," Mary said.

"Okay, well any time we can grow something on the moon we can save a whole bunch of time and money by not taking it from Earthside. You already know that. So we hope to set up biospheres and closed environmental life support systems several steps up from what we are doing here on *Friendship*. We want to see how plants perform in one-sixth gravity. We also intend to set up some one-third 'g' sections at our lunar ag station, too, so that we can begin to get a better understanding of what we might be able to do at our Martian base once it becomes operational.

"There was quite a lot of experimental work done with lunar soil after the Apollo crews came back from the moon last century. One that always stuck in my mind was the popcorn plant study done at Colorado State University down in Fort Collins. They used ten grams of lunar soil brought back by the *Apollo 16* crew from the lunar highlands at Descartes. They figured it cost them $1 million a gram to bring it back. Anyway, they planted the popcorn plants in three different soils for the test—the lunar soil, then volcanic basalt from Hawaii, and finally rich Iowa farm soil.

"After two years of study they decided that the lunar soil was no better than Earth dirt even though some of the scientists thought that the popcorn did a little better in the moon dirt because of the traces of iron, zinc, manganese, and copper that they found in it. At any rate, it proved that we can grow stuff up there in the native moon dust.

"Another thing that lots of people are speculating on is whether we should have poultry and cattle production at the ag station. It would do a lot for the morale of the crews on semipermanent duty up there to have fresh meat. So there are some very interesting experiments going on right now at Purdue University down in Indiana involving egg-laying problems in one-sixth gravity and in cattle breeding over at the University of Nebraska. Whether the pastures of the Midwest can be taken up to the moon is still too complicated to call. But at the very least we could handle the thing with artificial insemination.

"So all sorts of exciting things might yet go on up at the ag station. And don't forget that another consideration is the long 14-day lunar nights. That would screw up the hens and their egg production. So we'll have to design artificial lighting into the system as well. Being an old farm boy myself, I find this whole thing fascinating."

"What about the mass driver?" Mary asked.

"That entire project is all ready to be set up. We've had over 20 years to perfect it down at Princeton ever since Dr. Gerard O'Neill first thought it up. A lot of the teams involved think it could become a multibillion-dollar moon industry before it's all done. One of the NASA bosses, I think it was Bob Frosch, said that if we put 100 tons of machinery on the moon to reproduce itself, after 20 years we could manufacture ten million tons of aluminum per year. He also figured that if we set up self-replicating machines at the mass driver base, within two years we could have 100 mass drivers launching some 100 thousand tons of lunar stuff a year."

"Wait a minute, wait a minute," Billy said. "You're getting ahead of me. Exactly what is a mass driver? All of this other stuff is too far out for me."

"Good question. A mass driver looks like a long tube, actually like a chunk of pipeline, about 175 yards long. It really should be called a lunar catapult or a slingshot. What it does, by way of electromagnetic motors, is to accelerate buckets full of moon material to an escape speed of 1.44 miles per second in less than one second.

"Just before they get to the end of the driver tube, the buckets are friction braked and the material inside them leaves on a trajectory into the lunar sky. It doesn't take too much for it to escape the one-sixth pull of the moon's gravity, and it gradually drifts out to one of the moon-Earth Lagrange spots where it's caught in a cylinder ten yards wide. We call it the Lodestar Glove. In this case it's trapped at L-2 about 37,800 miles farther out from the moon on the far side from Earth."

"Wow!" Billy exclaimed.

"Yes, it's pretty amazing. But we're 99 percent sure it will work. Once we get our pilot driver set up we'll be able to prove it out for certain. And then we'll be off and running. We'll be able to mine the moon and transfer the material up to L 2, then take it to our low Earth orbit processing stations. We figure we can move about 1,800 tons of the material a year with just one such unit. L-2 was selected because the mass catcher can stay there permanently with a minimum cost in fuel."

"What is this L-2?"

"That's one of the five spots where the gravity of the Earth is cancelled out by the gravity of the moon. L-4 and L-5 are stable points where we won't have to use a lot of propellants to stay on station. They're located on the orbit of the moon around the Earth, one 60 degrees ahead of the moon, one 60 degrees behind. Both spots are about 233,000 miles from the moon. Then L-3 is directly opposite Earth from L-2. In other words, both the moon and Earth are located on a straight line drawn between the L-2 and L-3 spots of the moon's orbit. L-1 is not on the moon's orbit; it's a spot between the moon and the Earth. Right now that looks like the best jumping-off place for our Mars missions."

"What does the L stand for?"

"Well, it can stand for libration point or for Lagrange, a French fellow who first calculated that they should exist out there somewhere. He was a physicist and mathematician who lived back during the American Revolution. In 1764 he won a prize from the Paris Academy of Sciences for his work on the libration of the moon. He won the prize four more times during his lifetime for space-based computations. He's somebody I'd like to meet someday, if such a thing's possible. I want to ask him a couple of things about his Jovian theories."

"What about the observatory you mentioned?"

"That will probably be located near Galilaei Crater out in the western part of the moon, in the Ocean of Storms. Galilei Galileo was the Italian who first really used the telescope along about 1610 and made so many important discoveries."

"You say the observatory will be on the western side of the moon?" Mary asked.

"Yes, out near where *Luna 7, 8,* and *9* all landed. Those Soviet probes back in the sixties gave us our first glimpses of the moon's surface back when it was difficult even to hit the moon, let alone land on it. Matter of fact, *Luna 9* sent back the very first pictures of the moon, on February 3, 1966.

"One of the advantages of having our observatory on the moon is that it will give us a full 14-day-long look at the universe during the lunar night. Of course it will be in sunlight for 14 days, too, but the tradeoff will be well worth it. Another advantage is in the clean atmosphere, or I should say almost total lack of atmosphere, in the lunar sky. We're still arguing among ourselves as to how much atmosphere exists, if any. But whatever there is, there sure isn't much, and the observatory will give us some very clear views of the universe.

"What I call the core base is the largest outpost we'll have in the foreseeable future. It'll contain our research and development park. That offers some exciting possibilities, what with the vacuum, low gravity, large temperature variation, long day/night cycle, and all the other things that make the moon so unique. We also plan to have our solar power sub-station located there to interface with the silicon processing, solar array manufacturing, and development operations. We're aiming for a 10,000-megawatt range. Dr. O'Neill once figured that with an initial investment of about $60 billion we could reach a market $400 billion in size each year.

"Our core station would also be the headquarters for our lunar map-ping and geology sections. Mars has actually been mapped in more detail than the moon, believe it or not. Right now we're looking at putting this core station in the Sea of Tranquility. It has historical significance and could feed the tourist business out to the *Apollo 11* landing site down the road. There are other reasons for putting it there, but I suppose our plan-ners are humanists at heart, too, and the history thing is a pretty strong driver. As long as I'm on that subject, you might also be interested in knowing about the historical monuments that are planned."

"Oh, yes, I hadn't thought about that angle to the moon," Mary said.

"If you'll remember your space history, between 1969 and 1972 we landed twelve Americans on the moon. Their total stay amounted to less than two weeks, but in those two weeks they collected more than 800 pounds of lunar rock and soil and walked and rode their lunar rovers all over the place. They covered over 59 miles of the surface of the moon. They also set up six scientific stations that operated for years afterward and sent back data. The laser reflectors that Neil Armstrong and Buzz Aldrin put up at Tranquility Base are still in use, as a matter of fact. During those days 24 Americans in all flew out to the moon; the rest didn't land.

"Then there are other historically significant spots on the moon, such as the *Luna* landing spots that I mentioned earlier. Quite a while ago the whole gang got together in the United Nations and developed a list of historic lunar places. Number one, of course, is Tranquility Base. And after that all of the Apollo sites will be called by the name of the lunar lander that still sits there, Intrepid Base, then Antares, Falcon, Orion, and Chal-lenger. In addition to designating them historic sites that are not to be disturbed, we'll position habitation outposts near each with caches of food, fuel, water, and oxygen. These will be used for periodic scientific base camps as well as emergency shelters in case of some sort of solar flare emergency or some type of equipment failure.

"When the sites are dedicated we'll beam the ceremonies live back to Earth for global television coverage. A few of the Apollo guys are still alive and we hope to include them in our Earthside part of the programs.

"The other outpost that is still in phase A studies is a far side radio telescope. Being on the far side of the moon it will be away from all of the radio clutter coming from Earth. We're planning a joint U.S./U.S.S.R. outpost for that facility with just occasional visits by a crew to calibrate, update, and repair the facility.

"Where it might be located we just don't know yet; we've still got a lot of survey work to do. Some sentimental favorites that have been proposed so far are near Jules Verne or Yuri Gagarin craters on the far side. But with all of the variations in the moon's gravity because of the mascons—the buried mass concentrations—we just don't have enough data yet to make a decision. The far side outpost has to be one that we can navigate to without too much of a problem."

"*Friendship, Molly Brown.*"

Barney pushed over to the comm station. "Go ahead, *Molly Brown. Friendship* reads you."

"Just wanted you to know, *Friendship,* that we're station keeping with our skirt now and are about to bed down for the night. We'll take care of the changeover in the morning after we get some rest. We've had a busy day!"

"Roger, Kathy. How's your comm Earthside now?"

"No problem now, Barn. Once we got up here where we could maneuver easier, we've been able to keep in touch. We'll end now and catch some winks if you don't have anything else."

"Okay, kids. See you on the next shift. Sleep tight."

8

The General Store_____

Mary Two Hawks slowly sipped her Morning Thunder tea through the slim plastic tube. Bringing a limited number of personal food items was one of the personal touches that NASA permitted nowadays. It made the transition to orbit all that much easier, and Mary had started her days with Morning Thunder tea for as long as she could remember.

She was alone in the wardroom. The others on the blue shift had already drifted off to their work stations. But Stu had told the three of them to meet here, and she was the first to arrive.

Mary welcomed the time alone. Everything had been happening so quickly since their arrival. So much new information was being thrown at them. She was exhausted and had a lingering headache.

They had covered a lot in their preflight training, but there just was no way anyone could remember all of it. This was not a forgiving place—not a place to slip up and make a mental error. Knowing that it would make her seem less than perfect in their eyes, she was afraid to reveal her fears to the others. And for so long she had been afraid to make a mistake, to do the wrong thing, to say the wrong word.

She knew that this journey would end all too soon, and she wanted to

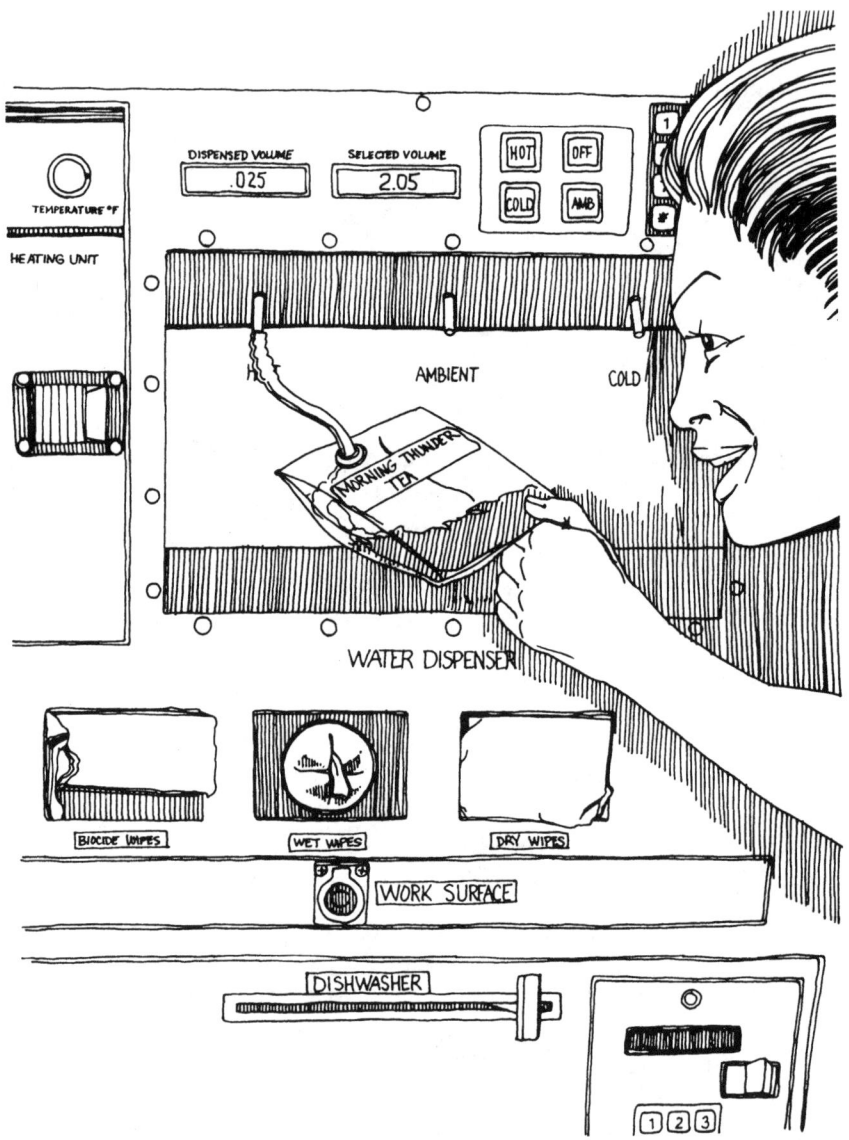

Mary Two Hawks enjoys waking up with her Morning Thunder tea. A limited number of personal preference food items are permitted for each crew member aboard *Friendship*.

relish and embrace each moment. She prayed to the Life Giver that she would not make any mistakes and that her thesis would be judged meaningful as a result of her observations.

Without even being aware of it Mary lightly touched the small buckskin medicine bag she wore around her neck. In it was the amulet that her great-grandfather had made for her out of a piece of cedar that had been struck by lightning. He had shaved it thin and made it into the form of the People. There was also a small, soft piece of doeskin with red and yellow crooked lines painted on it; her great-grandfather had told her it was big medicine. A small, clear quartz crystal and four eagle down feathers filled out the rest of the tribal pouch.

Mary suddenly realized that this groaning ungainly lodgepole of a space station hurried around the Earth to the east just as the tipis of her ancestors had faced east to greet the morning sun. "No," Mary Two Hawks thought as she sipped her tea high above the turning Earth. "It does not seem inconsistent to carry these tokens with me. Perhaps one day we will bend down and pick up a spent feather of another planet's eagle. And the faith of the People will once more be confirmed."

Mary was aware that the others had come through the node and were fixing themselves coffee at the water dispenser.

"Good morning, Sleepy-eyes. How did you sleep?" she asked them.

Billy Wong just shook his head, but Wayne Morrison managed a grumpy response. "Oh, all right, I guess, once I finally got to sleep. This sure is a noisy place to try to get some sleep. Lordy, think what it would be like if they didn't put so many sound deadeners in the bunkhouse."

"What's on the agenda for today?" Billy asked, joining Mary at the wardroom table. "Or is it today?" he said, as the great station slowly slid into night on the dark side of the planet. "This is really taking some getting used to, having the sun set when I get up in the morning or popping back up when I get ready to go to bed at night. Is it bothering you guys, too?"

Just then Stu stuck his head around the divider wall that separated the wardroom from the station operation area at the end of the module.

"Morning, gang, is everybody bright-eyed and bushy-tailed this morning?"

Smiles and groans in response.

"We figured you might have a letdown this morning now that you're getting settled into a routine, so Smokey and I are going to give you an easy day. Anybody want to help us on the log mod changeover?"

"Hey, yeah," Billy said. "That sounds like fun. What can we do?"

"Well, first of all, you can just watch while we do the physical transfer, but then you can do the manifest double check for us once we get that out of the way. Are you game?"

"Can do. What's first?" Billy asked.

"Smokey and I just about have everything set up between *Discovery* and *Friendship*. He has the bird powered up, the computer checked out, and the remote manipulator in idle. He was about to release the shoulder brace when I left to come get you. Stow your drinks or bring 'em with you and let's get to work."

"Damn," Wayne said as he floated into the far bulkhead. "I still can't get this business down right."

"Just slow down; you're in too big a hurry," Mary answered with a laugh.

"I guess."

Reaching the work station where Stu had already put his headphones back on, they formed a semicircle around him and watched over his shoulder.

"Here," Stu said. "I'll turn the speaker on so you can hear Smokey and Bob while we do this. Like everything else up here, we do it slowly and carefully. There's no need to be in a hurry."

"Confirm cargo bay lights on." Smokey's voice came to them over the speaker.

"Lights on," Stu answered, his checklist in his hand.

"Arm latch release," Smokey confirmed.

"Roger," Stu answered.

"Elbow and wrist cameras on."

"Check."

"Shoulder pitch on."

"TV downlink, enable."

"Thanks, Bob."

With *Discovery* still docked to *Friendship,* the two astronauts were standing at the aft crew station on the flight deck looking out into the cargo bay. Smokey stood on the left side at the mission station where he could monitor *Discovery's* systems and assist Bob, if necessary. From time to time he glanced over his right shoulder out the aft window at the logistics module sitting quietly in the bay.

Bob was at the payload handler's station with the remote manipulator translational hand controller to his left, the rotational one to his right side. The closed circuit television monitors were to his right at eye level next to the cargo bay window. He saw what the eyes on the elbow and the wrist of the manipulator arm saw in these monitors. He could zoom in or out with the cameras, change light levels, and focus.

Setting the X, Y, and Z axis translation coordinates and the pitch, yaw, and roll for the arm, Bob slowly moved the end effector toward the grapple probe on the logistics module. When he gently eased the three wires on the palm of the arm's hand over the grapple probe sticking out of the LM, the end effector ring began to rotate and the wires closed on the payload grapple, centering it and capturing the logistics module. Bob now held the LM in his mechanical fist, as he turned the mode switch to the orb-loaded position.

"Okay, Smokey, let's cut the lump loose from the nest."

"Hold-down latches released."

"Confirm."

The longeron and keel fittings and interface trunnions were now free of the orbiter. And the fully loaded logistics module, 24 feet long and 33,000 pounds, hung lightly within the belly of *Discovery*, where it was held tightly by the grapple capture.

"Well, that went smoothly enough, troops," Stu said to the watching rookies. "Now all we have to do is wait for dawn. I already captured the station LM before I came to get you. The mobile service center arm is all programmed to do its thing just as soon as Bob moves the fresh LM out of the nest. I've severed the fluids, power, and data lines already and verified the pressure integrity of the node. Our station arm will load the old module into *Discovery* for return to Earth, just as soon as we attach the new one with your clean underwear in it to *Friendship*. But we do have to do it on the bright side just in case anything goes wrong, and we have to do an EVA."

The station arm's reach envelope extended crosswise along the boom and lengthwise up and down the keels. There were actually two arms located on the movable mobile servicing center built by the Canadian Space Division. They were on either side of an EVA work station that could be manned by an astronaut when the work warranted one. The entire system was reminiscent of an old railroad handcar that ran up and down and back and forth on the leading edge of the space station as it leaned forward on its continual fall toward the waiting Earth.

As *Friendship* came out of the dark side of the planet into the bright glare of the terrestrial star, Smokey watched his panel as Bob slowly eased the freshly stocked logistics module out of the orbiter's cargo bay. Bob could see the swirl of clouds, continents, and oceans under the LM as he maneuvered it carefully away from *Discovery* and off to the side under the mainframe of the cluster of *Friendship*'s modules.

"All clear, Stu," Bob finally said, when he was sure that he held the LM far enough out at arm's length to permit Stu to do his work.

"Thanks, Bob. Confirm clearance and will initiate spent LM transfer."

The rookies watched the television monitors and looked through the windows as Stu slowly eased the now-separated logistics module with its load of garbage, human waste, freshly grown crystals, pharmaceuticals, and other goodies away from under the cluster of modules. Finally he commanded the mobile service center to carry the LM across the boom out of the way until the fresh LM could be put firmly into place.

"Okay, Bob. The berthing port is all yours. Come on in."

Inside the orbiter Bob entered the coordinates for the target into the computer and slowly moved the new logistics module horizontally toward the correct axis under the node. Finally he had it oriented at the berthing port, and he held it there until all of the vibrations had damped out. Inside the node another crew member stood watching the capture readout display, and when all was quieted down Bob eased the LM into the port.

After all of the connections for fluids, data, and power were confirmed and the new integrity was established, Bob released the end effector and returned *Discovery's* arm to its cradle.

Stu then ran the spent LM back over the boom to the orbiter's location and gently laid it in *Discovery's* cargo bay. When capture was complete, Stu released his LM and moved the MSC over to its storage area out of the way.

"Well, that was a morning and a half's worth of work, wasn't it, gang? What say we break for lunch? Smokey, we'll put the coffee on while you guys power down. Billy, see what you can find to eat in the galley. Cream of mushroom soup and a turkey sandwich sounds good to me if you can find one. And Wayne, you can fix the coffee. Mary, you can get the new LM manifest for me; it's over on my clipboard in the library."

After lunch Stu went over the manifest with the rookies. Each was assigned a part of the unloading and checking. Most of the consumable items were to be left in the general store for replenishing later.

"Now remember," Stu said, "I'll be looking over your shoulder, so don't be afraid to ask questions as you go along. You can cross off the atmospheric replenishment gases on the checklist now. I'll take care of those later, especially the liquid oxygen and nitrogen, and the lithium hydroxide canisters. We also won't concern ourselves today with the propellants for attitude control, reboost, and for the OMV and OTV. They're all out back on the LM pallet. We'll take care of those tomorrow. Now, any last-minute questions?"

"Yes," Wayne answered. "We're not to actually touch the new experiment racks. I understand that. We're just to confirm on the manifest that they are on board so that the science crew can transfer them later to the lab modules. Right?"

"That's correct."

"Okay, but what about retrofit kits."

"Just note that they're in the proper location. That will be enough."

"Just wanted to make sure. Well, gang, what say we get started."

The group made their way through the node and into the logistics module. It was 24 by 13 feet, but it looked larger to them up here in orbit than it had on the ground. In orbit, volumes such as this module often tended to look larger because a person inside it could move in all directions and did not have to look at a room only from floor level.

"Plenty of elbow room for all of us in here," Stu said with a chuckle. "Now, just to recap, we have 224 storage containers in here and 8 food containers. Also a 60-cubic-foot freezer—that's almost twice as large as Sears' best—and a 20-cubic-foot refrigerator—about the same size as the ones in your house, except this one is all refrigerator. We have 74 kinds of

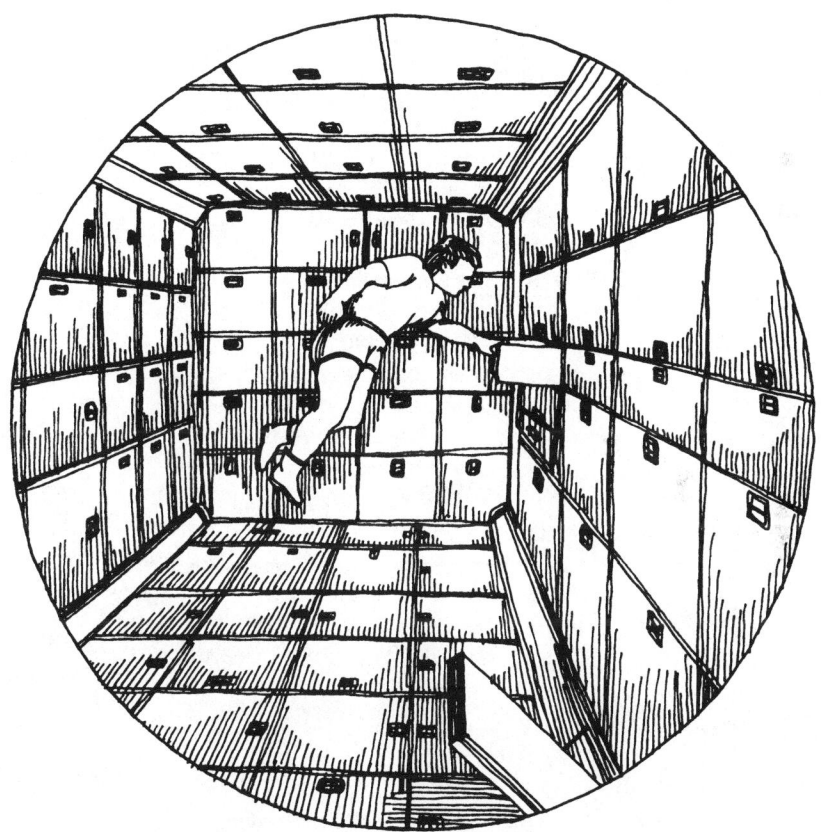

The interior of the logistics module. This 24 by 13-foot module is taken up fresh for every 90-day tour. It contains 224 storage compartments, a 60-cubic-foot freezer, a 20-cubic-foot refrigerator, 74 kinds of food, and 20 different beverages. It also carries up raw materials for processing, oxygen, consumables such as film and writing paper, and a wide range of scientific and space station equipment.

food on board and 20 different beverages, including Mr. Budweiser's best. It's a full 90 days worth of food."

Stu looked up at the storage drawers in the center area of the module. "Oh, yes. We also have a 90-day supply of bedding and a 14-day supply of clean clothing. Over there are the batteries and writing supplies. Here are the resupply items for the health management facility, and over there are the things for the Maintenance work station—people keep stealing screwdrivers up here. The housekeeping things are in here, and down there are the safe haven supplies.

"Now this is one of the areas you should not open. In this rack are the raw and processed materials the commercial crew uses in their daily operations, and nobody touches that stuff for proprietary reasons. 'Nuff said. But speaking of that, up here are the experiment supplies, such as the gases, film, and the like. That area is a no-no, too. Those people will check their own areas later this evening."

"What's up there?" Billy asked.

"Those are the spare parts and panels as well as some special maintenance and servicing equipment that we had you bring up this time. Oh, and I almost forgot the good part. In here are your books, magazines, the new movies, and some fresh music cassettes we ordered.

"This drawer has gum, candy, breath mints, peanuts, and cookies. I'd better warn any cookie monsters among you—make 'em last. Okay, that about does it. I'm sure I forgot to point out some general areas to you, but as I said, I'll be in the area if you have any questions. Mary, why don't you and Wayne start over here with the food lists while Billy gives me a hand with the payload supplies and spares."

"Will do."

Mary and Wayne floated over to the food storage area back near the hatch. "Man, don't those fresh vegetables smell good," Wayne said near the greengrocer's bins. "Am I ever glad sweet corn and tomatoes were in season. And those cantaloupes smell like they're almost ripe. Does anybody know if they loaded any Vidalia onions on board?"

"You and your onions. Now how about you checking things off as I read them to you?"

"Fair enough."

"Let's see, this container says Beverages. Apple drink; cocoa; coffee-black; coffee with cream; coffee with cream and sugar; coffee with sugar; grape drink. Instant Breakfast. Lemonade; orange drink; orange-grapefruit drink; orange-pineapple drink; strawberry drink. Tea. Glad I packed my own Morning Thunder tea. Ah, let's see, where was I? Grapefruit drink; more Instant Breakfast, this time chocolate and strawberry; tropical punch; and more tea, this time with lemon and sugar. Got that?"

"Per the list. So far so good."

"Next container says Cereal. There's bran flakes; cornflakes; granola with blueberries; and granola with raisins. What, no oatmeal? What would my grandmother say? Next we have fruits. Let's see, applesauce; apricots-dried; bananas-freeze dried; fruit cocktail; peach ambrosia; peaches-dried; peaches-thermostabilized; pears-freeze dried; and pears-thermostabilized. Let's not forget the pineapple-crushed and the rehydratable strawberries."

"Does fruitcake go with fruits or cookies?"

"Who knows? Now the meats. Beef almondine; beef-corned; beef and gravy; beef-ground with pickle sauce; beef jerky—ah, my ancestors would be proud. Next, beef patties; beef-slices with barbecue sauce; beef steak-thermostabilized and irradiated; beef stroganoff with noodles—"

"The famous mystery dish, you know."

"Right. Next, chicken à la king; chicken and noodles; chicken and rice; chili mac with beef; frankfurters; ham; meatballs with barbecue sauce. Did you know that the guys on *Apollo 12* kept kidding Al Bean about how much spaghetti and meatballs he could put away? Now, why would something like that pop into my mind? I must read too much. Next, salmon; sausage patties—hope we have plenty of scrambled eggs some-place—shrimp creole; and here's spaghetti; then tuna; turkey and gravy; turkey smoked; and finally, turkey tetrazzini."

Mary and Wayne continued on down the checklist as instructed, never thinking that the on-board computer already had all this in its inventory control system. This exercise was part of a planned psychological indoctrination period for them. Among other things, it was to make them feel at home and to allay their fears that they were completely separated from the Earth. One day before the end of their tour it would no doubt occur to them all—although like most others before them they would not likely say anything about it.

The three rookies also did not know that Stu was keeping a careful eye on them this very minute as they continued what they considered to be a very routine task inside the logistics module.

Long before they had been assigned to this tour of duty aboard *Friendship*, the three had been watched, and finally they'd been selected to be part of a long-range heterogeneity study for NASA's human factors office. The purpose of the study was to determine the effects that ethnic background, age, gender, and several other factors have on very long space flights. By studying various groupings on these tours, NASA was slowly building up a profile of an optimum crew composition for multi-year deep-space flights.

How people work together doing routine tasks such as this pseudo inventory tells much about how they would respond together in a catastrophic event.

How mixed groups like these rookies respond to privacy invasions, crowding, and similar factors is also important. So too is the proper mix of cultural backgrounds as crew sizes increase. Conflict, as well as communications, performance, and morale, were also being studied among the rookies.

Mary Two Hawks, Wayne Morrison, and Billy Wong would probably never know that they were part of this soft research, yet their contribution to the future of mankind in space would go far beyond their preflight goals.

"Next are the snacks," Mary said to Wayne. "Ah, let's see. There's Life Savers; cookies-pecan; cookies-shortbread; crackers-graham; food bars-almond crunch, chocolate chip; granola; granola with raisins. Then we have nuts—almonds, cashews, and peanuts. Where do we classify peanut butter?"

"I'm not sure, but I sure wish we could have brought fresh bananas along. I love them smeared with peanut butter."

"Yuck! You got weird taste buds, kiddo."

9

The Industrial Park_____

"You all met Edie White last night at supper, right?" Stu asked, as he and the rookies floated into the materials processing module.

"Yes, we did," Wayne answered for the group. "And I for one am very interested in spending some time today with Dr. White."

"Well, good," Stu answered, "because I'm going to leave you here with her while I take care of some business. See you all later."

"Well, people, it looks like you're on your own with me for a while," Edie said as Stu floated through the node and turned into a hatchway. "Let's see what we can think of to help you understand what we do here." Motioning to the rookies to follow, Edie began to move around the module. "It's such exciting work," she said. "Our research is constantly showing us new and better means of production that can be used on Earth.

"In materials processing we're concerned with four major areas: crystal growth, containerless processing, fluid and chemical processing, and bioprocessing. We're doing things up here that we simply can't do on the ground because of gravity.

"We're doing so much up here that it's really hard to know where to begin. It's very important, though, that you remember that we're not working in the blind up here. I was going to say that we're not working in a vacuum, but that's too thorny to speculate on.

"Virtually all of our day-to-day production and experimental work is directly tied in to mutual support units on the ground. We're part of the Universities Space Research Association network—USRA, as we call it. It's a consortium of over 50 American universities. There are three different regional USRA institutes around the country right now. There's the Lunar and Planetary Institute in Houston; the Institute for Computer Application in Science and Engineering at the NASA/Langley Center in Virginia; and the Research Institute for Advanced Computer Science out at Ames in California.

"The whole USRA thing started at Lehigh University in 1970, the year I was born. And we're all tied together in one homogeneous computer network between *Friendship* and all the Earth laboratories with our numerical aerodynamic simulation supercomputer system at Ames. We're using parallel processing between our old Cray-2s and the later high-speed processor 2s. Right now we can perform 10 billion computations a second with the NAS system. The various centers tie into the intellect with the UNIX system, and our ADA-based system is driven by the NAS as well.

"So we support their research on the ground and they support our work up here. That enables us to do a lot of the 'quick is beautiful' science that Freeman Dyson alerted us to back before the station was built. He said something to the effect that the most important discoveries are those that cannot be planned in advance and that if we want to do good science in space we must be able to jump quickly to take advantage of unexpected opportunities. Kind of an instantaneous sharing of the truth as we know it up to this moment.

"We're doing joint projects that we couldn't even believe were possible back when I was in school. We're making glass that bends without breaking; we call it metallic glass. It's actually a metal that ends up with an amorphous structure like glass when it is rapidly cooled. Caltech discovered it almost 40 years ago, and we're using it now in electrical transformers and saving 75 percent of the energy usually lost in switching electric currents. That alone saves us over $1 billion a year. Our research up here is extending the envelope for that technology.

"In our polymer research, we're improving on plastics that are already stronger than steel and lighter than aluminum. And we're even using a bacterium technique in our biopolymer section to produce a better grade of biodegradable biopol for use in bone replacement surgery. That's a British development. The biopol functions while the bone mends, and then it degrades and eliminates the need for a second operation to remove the plate, as we had to do in the old days.

"In our composite materials research we're reinforcing ceramics and plastics with carbon filaments and synthetic fibers. So far we've been able to replace a lot of the metal parts in aircraft and automobiles with lighter-weight composites and saved a great deal in fuel as well as tooling costs. Here on *Friendship* we're developing microgravity blends of the various materials in our shopping bag and searching for newer, stronger, lighter composites.

"We're finishing up a project right now on the directional solidification of magnetic composites to follow up on some original work that was done by Grumman many years ago. We recently got some new ideas about it. And our work for the automotive industry continues for ceramic engine components.

"The Soviets made a breakthrough in semiconductor chip production when they discovered a way to deposit an ultrathin diamond film under the chips. This helps dissipate the heat buildup better. Our people in the United States are combining this with our ongoing gallium-arsenide research to build even faster chips."

Edie and the rookies had moved to the furnace and containerless processing module. She explained that the state of the art was fairly small scale compared to what was on the drawing boards for the next 20 years. Since the sixties scientists had tried to do some of the experiments in drop tubes, in aircraft, and in spar rockets. But they were really limited as far as what they could actually do. The drop tube experiments gave them only three to five seconds of low gravity, and when they switched to the KC-135 aircraft flying parabolic trajectories, they could stretch it out to only 30 to 60 seconds.

With the spar rockets they found they could achieve microgravity for about five minutes, but that still wasn't long enough. They did manage some MPS experiments on *Apollo 14,* including fluid electrophoresis, heat flow and convection, composite casting, and liquid transfer. Some experiments were also done on *Apollo 16* and *17;* but an equipment problem ruled out *Apollo 15.*

On *Skylab* they were able to do 16 different materials processing experiments, some of which lasted up to three months. Another dozen experiments were completed on the *Apollo-Soyuz* flight with the Soviets in 1975. The shuttle opened up many more opportunities, and in 1982 they started electrophoresis work in earnest as well as work on latex spheres.

The spheres were the first made-in-space products. In 1985 the National Bureau of Standards started selling them for $400 a vial as calibration tools for scientific instruments as well as for measuring things like red blood cells. The spheres are so tiny that they were first sold in 5-milliliter vials that held about 15 million of them suspended in water.

Other uses for the spheres include measuring the size of pores in the walls of the intestines in cancer research and measuring the pores of the human eye in glaucoma research. A lot of the work on the spheres was done by Dr. John Vanderhoff at Lehigh University, Edie White's alma mater.

Edie went on to say that an updated version of the monodisperse latex reactor, or MLR, was out on *Free Flyer 2.*

"That's old number 2 on a tether right out there trailing alongside us," Edie said. "We've got a lot of our more sensitive things on board the free flyers to keep them away from the tiny vibrations we have here on board. We can reel them in any time we want to check on things. At any rate, the MLR ovens are out there, and once we seed them, they're on their own. The spheres are actually made of polystyrene; that's a plastic of sorts that we do some special things with using carbon molecules and a benzene ring. But you'd better not let me get into all of the nitty-gritty of all these things or you'll be here all week. Does anybody have any questions about the microspheres?"

"Exactly how big are they?" Wayne asked.

"Well, you can't see them with the naked eye, if that's what you're asking. But, under an electron microscope they look just like billiard balls. And that's the secret to why they are so valuable compared to the same sort of thing we tried to produce Earthside. Our *Friendship* spheres are all uniform. We got into some real gravity problems down below when we tried to go too large with them.

"Up here we usually stick to from 10 to 100 micrometers; that's where the demand is, and remember, this is strictly a business proposition. We do our share of R&D work up here, but eventually the bottom line rules whether we keep on with an experiment or not. There's only so much seed money available, and we have to follow our business plan."

"You mentioned Lehigh, Dr. White. Is that where you got your doctorate from?" Mary Two Hawks asked.

"Oh, no. I just did my first four years there. I wanted to get off on my own after that and so I headed up north to Cornell. That's where I got involved with all of this gallium-arsenide business.

"You know, silicon is a very common material on Earth; matter of fact, next to oxygen, it's the second most common element there is in the Earth's crust. And so when our computer technology was born, it seemed rather appropriate that silicon would end up being the basis for all of our microcircuits. For a long time all of our computer chips were built on a silicon base. But it had certain limitations to it that we had to solve if we were going to make major breakthroughs into supercomputers and other areas where superspeed was the driving design factor.

"People like Harry Gatos at MIT kept working on the problem and eventually came up with gallium as one solution. It's a lot more rare than silicon. Gallium is a byproduct of aluminum refining. Arsenide is a by-product of lead and copper refining, and the combination of the two resembles table salt. It really complements silicon rather than replacing it as so many people thought at first.

"Gallium-arsenide has a lot of advantages and has given us a jump in things like advanced communications systems, high-speed data processing, artificial intelligence, 'smart' weapons, and microwave signal processing. But it costs more than 70 times the price of gold. The last time I paid any attention to it, it was well over $30,000 an ounce. It's a very brittle material and it's difficult to work with, which leads to very low yield rates when it's being produced. The end cost is about 100 times more than silicon-based microcircuits.

"Another disadvantage is that it requires metal gates instead of the oxide gates that we use on silicon chips. But people at places like the University of Illinois kept working at it and helped a great deal in coming up with epitaxial layers of gallium-arsenide on silicon substrates. That combination is really growing in popularity.

"By itself, it's difficult to beat G/A if cost is not a factor. It moves electrons three to six times faster than silicon, emits light where silicon doesn't, absorbs sunlight more efficiently, and uses less power. Then too, it is more resistant to radiation from the electrons from the sun and from possible nuclear bursts. And, since it can operate at higher temperatures—silicon fails at 200 degrees Celsius—it reduces the cooling requirements for computers and other electronic systems.

"G/A can process both light and electronic data on a single chip. Hewlett-Packard developed a G/A chip that transmits information at 5 billion bits per second. That's pretty tough to beat for efficiency—although indium phosphide is even faster, but that technology is still under development.

"High resistivity is also a big plus with G/A. That means we can isolate electrically adjacent devices on a G/A chip without using a sapphire substrate as we have to do with our silicon chips. All in all, G/A makes computer chips smarter and about five times faster, more powerful, and less expensive per computation than silicon."

"You've got me convinced," Wayne interrupted; "where do I buy stock in your gallium-arsenide company?"

"Sorry, I can't help you with that end of it; all I do is help make the crystals up here. This is the furnace over here, by the way. Our new, large commercial size. We just replaced our old one a couple of flights ago. We've got all of the bugs worked out now, hopefully, and are on stream with it.

"Basically, what we do up here in microgravity is build up a G/A crystal seed layer by layer, using an electrical current. Since we can avoid the convective stirring that normally occurs on Earth when the melt is heated, we get higher growth rates. And, since we can grow the crystals by using this Gatos method at lower temperatures, the material is more stable and we can hold down impurities. It also eliminates the dangerous arsenic vapor problem that we'd get otherwise."

"How big do these crystals get? How much can you grow in a year?" Billy asked.

"I'm sorry, Billy, but that's something that is proprietary and that I just can't talk about. I can tell you, though, that it's a $140 billion market, so we keep the furnace running at peak capacity with very little planned down time."

"Is it all used for computers?" Mary asked.

"Oh, no. A lot of the things that you take for granted nowadays are actually in existence thanks to gallium-arsenide. Of course, all of the LED displays in your watches and pocket computers are G/A and so are the collision warning systems in your automobiles. You can thank gallium-arsenide for the small rooftop dish antennas that you have at home. In the olden days, people used to have huge dish antennas in their backyards where they couldn't get satellite transmissions by cable. Now most everyone has a small dish antenna to pick up the signals.

"Let's see, what else? Well, the whole supercomputer industry is built on gallium-arsenide and the gallium-silicon technology that I mentioned earlier. Oh, yes. Your Dick Tracy wristwatch radio-telephones that keep us in touch with our loved ones use G/A. They've been a real godsend for letting people on the go stay in touch with home. To say nothing of the law enforcement, medical, and military uses. They all receive and transmit off GEO satellites, just in case you didn't know, and it doesn't matter how far out in the woods you are.

"What else? Oh, yes, the inexpensive car telephones that we all have on our dashboards now use G/A chips and so do a lot of the medical instruments we use in our hospitals and doctors' offices. Also the smarter and faster robots that we have in our factories and the high-speed seismic processors that our people use in the field. And our aircraft and shipboard satellite navigation systems."

"You mentioned 'smart' weapons awhile ago," Billy said. "Exactly what did you mean?"

"That's another area that I have to be careful of, but I'll tell you what I can. I think you were all briefed on our Strategic Defense Initiative, our Star Wars system, weren't you?"

"They showed us a videotape during our preflight training, but that's about all," Mary answered.

"Well, the G/A that we make up here is used out there on our Star Wars shield. Our SDI spacecraft are controlled by G/A computers, and our Stealth-type G/A evasive radar systems are also part of the system. It takes an enemy radar signal and plays games with it before it lets it return. This all happens quicker than a snit, you know. At any rate, it's all a part of our MIMIC program down at the Department of Defense. And, our high electron mobility transistor technology is also getting a lot of use by the Air Force. It provides switching speeds of ten trillionths of a second.

"Speaking of that, our E-lab worked on part of the high-speed data processors when I was in school. They came up with tiny laser beams to replace the wire interconnections between printed circuit boards and microcircuit chips. This all made computing by light a reality. The lasers are made of G/A and are the size of a grain of salt. And are they ever durable. They'll easily last for 100 years. A scientist by the name of Alfred Cho at Bell Laboratories came up with the molecular beam epitaxy technique that we use. This fiberoptics system uses hair-thin fibers at billions-a-second binary 'yes-no' rates. It's really something when you dig into it."

"I think my own circuits are getting overloaded with all of this," Wayne admitted. "It's an awful lot to try to absorb all at one time."

"I know it is," Edie said, "but I've covered just about all of the basics, I think, with the exception of the G/A solar cells. I mentioned that G/A can operate at higher temperatures than silicon and is more resistant to radiation damage from the sun. It also absorbs more solar radiation and is more resistant to heat. This permits our energy collectors to operate at GEO whereas the old-style silicon-based solar cells gave us fits at those higher orbits.

"The gallium-arsenide solar cells are more efficient, too, with about a 21 percent efficiency in converting solar energy to electricity compared to 14 percent with silicon cells. And, in three to seven years the G/A cells still retain 90 percent of their original power, whereas a silicon cell deteriorates down to about 60 percent. G/A gives us thinner and lighter solar cells as well as increased per cell power output. This means we can boost the power available for payloads or reduce the area of the solar arrays by using fewer cells. Our solar arrays right out there on the ends of the truss use G/A cells, by the way.

"Well, I think that about covers it," Edie said, "Unless any of you have any questions?"

"You've got to be kidding," Wayne said, laughing. "We're not sure we even absorbed all of that, let alone think up new questions to ask. Whew!"

"Okay. Well, then, let's all move down here and I'll cover containerless processing. That's not quite so involved, and I think it'll give your minds a rest."

"Sounds good."

"Maybe I should cover a couple of other aspects of this materials processing business before we get into our containerless equipment. A few minutes ago I mentioned the R&D work that we do. Now, most people think that our R&D work up here is just to discover ways that we can produce things here on *Friendship* that we can't make Earthside. But that's only partially true. We also are trying to create new techniques and materials up here that we can later try to duplicate down on Earth, in the gravity environment. It just costs so darn much to bring up the raw materials in many cases that it's simply not cost effective to try to do the follow-up manufacturing up here.

"Yet we've stumbled on several important techniques that we've already figured out how to duplicate down below. One is the foam steel that's so light that it floats on water. We did the initial R&D on that up here. And right now we're perfecting our low-density aluminum techniques. That will really save a lot of weight.

"The front-end costs in both time and money are the crucial factors. If proof-of-concept requires a long lead time before a payback, then many companies are reluctant to take the technical and financial risks of a 'reach' venture like we do up here. Some of the companies that showed corporate foresight back in the early days of our experiments on the old Spacelabs were 3M, McDonnell Douglas, John Deere, and Honeywell. Now they're all familiar to you because of the products that they offer based on space technology, but back in the old days they were really taking a calculated venture-capital risk.

"At first we were limited to processing samples of high value, low volume, because of the space and weight restrictions on the orbiters. Since *Friendship* was created, we've been able to greatly expand our production capability. That's had a dramatic effect on the release of more venture capital from private industry and the creation of more consortiums.

"An important thing to remember is that we're researching up here on how to do things in space so that once we get our fleet of space voyagers headed out toward the far shore, we'll be able to produce things we need as we travel. Some things will be produced from materials we encounter along the way from captured asteroids or materials we gather during drops down to friendly planets.

"Somehow I got off the track. This is our containerless furnace corner. We do our crucible-free processing here. The fact that we do not use a vessel means that we can work contamination-free. We get smoother surfaces and no nucleation. Using several techniques we suspend the molten materials until they can cool and harden. We have a three-axis levitator

Edie White explains some of the materials processing equipment. Most of the work on board is done in conjunction with scientific and industrial teams on the ground. The real-time exchange of information is vital to the success of this module, and many processes are televised for the benefit of those watching from the ground stations.

developed at JPL that works acoustically, and we also have electromagnetic and electrostatic facilities.

"We can form these molten blobs to shapes to meet contractor specifications by the use of sound waves. They hang together because of surface tension of the material. On Earth, when we mix two materials the heavier one sinks down to the bottom as the mixture cools. Up here we can blend new types of materials because of the lack of gravity and convection flows. Then, when they cool and harden, we get a new material to experiment with.

"Glass processing is one of my favorite departments here in this module. We're doing so much that we could only dream about just a generation ago. We're even using very deep undercooling, so we can use some materials for glass formation that we couldn't even think of using 20 years ago. Obviously, the implications of that on our limited natural resources on Earth are significant, to say nothing of the possibilities it opens up for using captured asteroid resources when we make the Big Trip.

"Optical glasses are one big growth area. We're really excited about our new batch materials using alkoxysilane, metal alkoxides, and metal salts. We expect to achieve some ultrapure opticals that way. And you know the purer the glass in fiberoptics, the farther the signals travel through the fiber before they disappear.

"Just to give you one example, contrary to popular opinion we do not use satellites for all of our telephone communications between continents. What we use for a lot of the traffic are fiberoptic cables, hair-thin strands of glass that conduct light rather than electricity. A tiny laser is used to transmit information, including voice, and just one of our new fibers can carry hundreds of thousands of conversations at the same time. Du Pont has done a great deal of work in this area with us.

"At first our gel-derived glass research was at an impasse because of the glass bubble problems we were having here in microgravity. What they call Stokes bubble rise just didn't occur up here in a predictable way, but the people at JPL did a lot of work on gas bubble behavior in multibubble systems and finally worked out our present techniques.

"I also mentioned a few minutes ago that we have the very latest version of our multiple-beam laser melting furnace with a triple-axis acoustic levitator. We can zap things and then do magic tricks with them just like an old circus juggler.

"Another area we're working in is the fabrication of deuterium-tritium reactor targets for use with Lawrence Livermore's NOVA laser down in California. And we've even supported some research on X-ray–driven targets, but that's still classified and I can't talk about it."

"Question, Dr. White," Billy interrupted. "I'm not sure I understand these targets you're talking about. Exactly what are they made out of?"

"Well, let's see. We have an inner gold-covered ball filled with a deuterium-tritium liquid mixture. That's surrounded by cold helium gas. The next large layer of the golden apple is more deuterium-tritium. Then CH polymer seeded with more gold, and finally an outer layer of lithium. We handle all of this coating and recoating in this levitator furnace over here for the inertial confinement fusion program. We're also working in this area on metallic glass spheres with JPL, too, for use as targets.

"Actually, this whole pellet or target area is changing rapidly as we close in on better techniques with different materials. We're making progress quite rapidly toward a workable nuclear fusion system, and we're still four years away from Gerard O'Neill's deadline. Way back in 1976 he said that we would most likely take another 35 years to get hydrogen fusion to work. Time's running out, but we're close to meeting that prediction.

"Whether or not nuclear fusion systems will ever replace fossil-fuel power plants is another question that we're trying to help the Earthside teams answer. The goal has been to get the cost of these reactor targets down in the 50-cent range. In order to get an economical energy-producing reactor, the target cost has to be low. They'll feed these darn things in at ten per second, day after day, year after year. So we're trying to help them decide the best way to mass produce them."

"Is this the same way," Mary asked, "that we're going to use nuclear fusion for spaceship propulsion?"

"Well, the people at Livermore and JPL are working up final plans right now on a 1,000-ton space freighter that will be able to haul cargo all over our solar system. We'll be able to zip out to Mars with it in just two weeks once it's working properly. That's compared to the 210 days outward bound that it took our unmanned sample return spacecraft in August 2003. Mars was at its closest approach to Earth then or the trip would have taken even longer. So we're really anxious to get our laser-induced pulse fusion ship on stream. As you may know, we're conducting a prototype field test of just such a vehicle in a day or so. It's called Project StarEagle. Let's move on over here.

"Down at this end of the module we're working on a long-term project with the people from John Deere, the farm implement company. We actually provide the on-board support for them and interface with their people on the ground. They've had a very active space-processing program since the mid-eighties, searching for new alloys and new crystalline structures. They got interested in space because of a fluke thing that happened. Have any of you ever been to the Field Museum on the south side of Chicago?"

"Yes," Mary answered, "I was there on my senior-class trip in high school."

"Well," Edie went on, "the museum asked Deere to saw some meteorites in half for them, and when the people at Deere found graphite nodules inside similar to what they have in the nodular iron products they produce, the light came on. They've been searching up here in microgravity for better ways of making engine blocks and other cast-iron parts. They're studying the microstructure of the iron, using a furnace developed for them by Grumman and Marshall.

"Deere's foundries make about 600 million pounds of cast iron every year, so if they can perfect the technique they're working on to produce more spherical graphite nodules up here, they'll greatly increase the strength of the iron. And the large diesel engines they use really have a stress load in all directions. Deere is hoping to lower their foundry operation costs and be able to produce more stress-resistant materials. So even something that most people take for granted like plain old cast iron can offer exciting research fodder here on *Friendship*.

"You know, a lot of the work we're doing now was really kicked off on the old ISFs, the industrial space facilities, that Maxime Faget's Space Industries in Houston put up three or four years before we started assembling *Friendship*. The ISF was a pure R&D laboratory up here that gave us a toehold.

"In case you didn't know, Faget was the man that everyone called the father of the Gemini, Apollo, and Skylab programs.

"The ISFs look a little like a giant dragonfly with their long solar arrays spread out on each side. We're due to haul one back into port next week, so you'll have a seat on the 50-yard line to watch us work on it.

"A lot of our work in this module is piped directly down to principal investigation teams on the ground. They can sometimes see things that we can't and give us advice. We use a lot of holographic interferograms and schlieren photography up here to help them see what effects microgravity has on the experiments we run."

Wayne held up his hand. "What is schlieren photography?"

"That's a technique that is several hundred years old now. As a matter of fact they used it way back in the nineteenth century to detect flaws in low-quality glass, so it's kind of ironic that we're still using it in glass production up here. Schlieren means 'streaks' in German. It's a technique to make transparent phenomena visible. For example, it reveals flow patterns during growth cycles and helps us study turbulence problems in heat flow. Back in the early days of the shuttle they used it to study shock waves around a model of the shuttle design at Ames Research Center's wind tunnel.

"Schlieren photography uses lenses and mirrors to project an image onto the film. Changes in light refraction are revealed with this kind of photography. It's useful to engineers and biologists alike. We use colored filters on our equipment up here to help us track things better.

"Well, I think that pretty well covers this module. If there aren't any questions let's move into the next module where we do our electrophoresis work. Incidentally, Wayne, we're looking forward to having you here to work with us for the next 30 days. I'm sure we'll both learn from one another."

Sunrise occurred just as they entered the life sciences factory module. Without realizing it, they had adjusted over the past couple of days to this frequent change from day into night and back.

In the last module the scientists were looking far into the depths of the materials they were working with in order to affect their molecular structure, characteristics, and potential uses. One of the most important tools they used in that area was the scanning tunneling microscope through which they could look at individual atoms. It was invented by a West German named Gerd Binning who worked for IBM in Zurich. Magnifying things millions of times, the microscope was extremely useful in the materials processing area, particularly in looking at integrated circuit coatings for the semiconductor industry.

The scanning tunneling microscope was also used by microbiologists to look at the molecules in human cells. So it was an essential instrument in the life sciences factory module as well.

In both of these modules they were attempting to learn the secrets of both the atom and the human gene. The physicists were studying the atom, and the molecular biologists were studying the gene. At times it seemed that they would discover that both animate and inanimate objects work in just about the same way.

"In this module," Edie explained, "we're trying to find out how the one-dimensional linear genetic sequence in the DNA genetic code produces human beings, a three-dimensional organism. Just how do our genes switch on and off, how do our individual cells communicate with one another? If we can begin to understand these things, and a great many others, perhaps we can learn to control birth defects and even regenerate living nerve tissue.

"That too is all a part of our drug research up here. We're looking for new information in basic cell biology so that we can develop new pharmaceutical approaches in a rational way. Millions of us will live longer and live more productive lives. Think of the symphonies that might yet be written.

"Our goal here on *Friendship* is to produce lifesaving medicines worth as much as $22 million a pound. We can make them about 800 times quicker and much purer than we can on the ground. Just one month's yield of drugs up here equals over 30 years' production of the same drugs on the ground. And not only that, but they are five times purer and that much more effective.

"And while it takes thousands and thousands of people to support this space station and all of its adjacent facilities, it really does still come down to the fact that within the mind of each of us on board at any given tour, the quiet, seeking work of discovery goes on. Each of us is the assembled knowledge and experience of one life, and each of us looks at a problem and its possible solution in just a little different way.

"I remember reading once that Robert Weinberg of MIT got his flash of insight about how this genetic code could trigger cancer while walking across the Longfellow Bridge in Boston during a swirling blizzard. He followed up on his idea and searched through the 100,000 genes in a human cell until he found the single mutant DNA nucleotide that caused bladder cancer.

"Research is not glamourous; it's tedious, plodding work. But discovery is magnificent, especially when it leads to the elimination of pain and suffering.

"We've also been collaborating with people like Jacqueline Barton at Columbia to try to isolate individual genes on the DNA helical strands to gain a better understanding of why proteins activate some genes and not others. Mapping these genes and discovering the sequence they operate in will help us understand our human bodies and the purpose of each gene. Learning how they operate in cancer and heart disease is one of the goals of our biotechnology. Someday soon we will be able to take a blood sample from a newborn baby and predict the diseases that that child will be susceptible to, based upon his genetic makeup. His parents will be able to adjust his life accordingly.

"This is really very detailed detective work that we're doing. Yes, discovery is magnificent and could affect every living creature on Earth, to say nothing of those living beings that we will encounter as we journey far out into the stars.

"Let's talk about specifics for a while. Exactly what drugs are we working to produce and how are they used? First, you have to understand that this is big business up here. McDonnell Douglas makes over $1 billion worth a year of just *one* of the drugs they manufacture up here. And they make quite a number of different drugs in this section of the station. So do many other companies, now that the technology is on stream. The total market value of the drugs we produce up here is now close to $30 billion a year.

"Let's see, urokinase is worth about $1,000 a dose and is used for

Medical Benefits of Space Station Research
Some Agents for Electrophoretic Processing in Space

Typical Agents	Beneficial Medical Applications	Function/Status	Annual Patients (USA)
α_1 Antitrypsin	Emphysema	Research quantities only now	100,000
Antihemophilic factors 8 and 9	Hemophilia	100% terminal by age 40	20,000
Beta cells	Diabetes mellitus	Possible single-dose cure	600,000
Epidermal growth factors	Burns	Replacement skin grafting	150,000
Erythropoietin	Anemia	Replacement transplants/transfusions	1,600,000
Immune factors	Viral infections	Treating immune disorders	185,000
Interferon	Viral infections	Potential may be unlimited	>10,000,000
Granulocyte stimulating factor	Wounds	Research quantities only now	2,000,000
Lymphocytes	Antibody production	Replace antibiotics/chemotherapy	600,000
Pituitary cells	Dwarfism	Currently not curable	850,000
Transfer factor	Leprosy/multiple sclerosis	Potential for other applications	550,000
Urokinase	Blood clots	Low development costs	1,000,000

Source: Reprinted by permission of the publisher from M. H. Harrison, RAF Institute of Aviation Medicine, "Space Station: Opportunities for the Life Sciences," Journal of the British Interplanetary Society, March 1987.

people with pulmonary embolism and those who have suffered heart attacks. Over a million people in America suffered from these diseases at any given moment back in the eighties, and 50,000 of them died from them each year. Our production capacity is about 500,000 doses a year.

"Factor 8 is another one. It's worth $3,000 a dose and is used in treating hemophiliacs.

"Beta cells normally produce insulin in the pancreas. We make them up here and by transplanting them into diabetics we have come up with a permanent cure for diabetes. This disease was the third largest killer in the United States, just behind cancer and heart disease. And not only that, but diabetes is an insidious disease that also leads to kidney failure, heart disease, and blindness. So beta cells can also help relieve human suffering in those areas.

"We're well on our way to accomplishing our long-range goal here in this module of discovering or developing 15 new medicines in a 10-year period. That was one of our criteria for success that we set for ourselves when we put the station up here. Without this manned module we feel that we could only have discovered three new biological materials in the same period of time on unmanned free flyers."

"What is going on in this unit, Dr. White?" Billy asked.

"That's the joint McDonnell Douglas/Lovelace production facility,

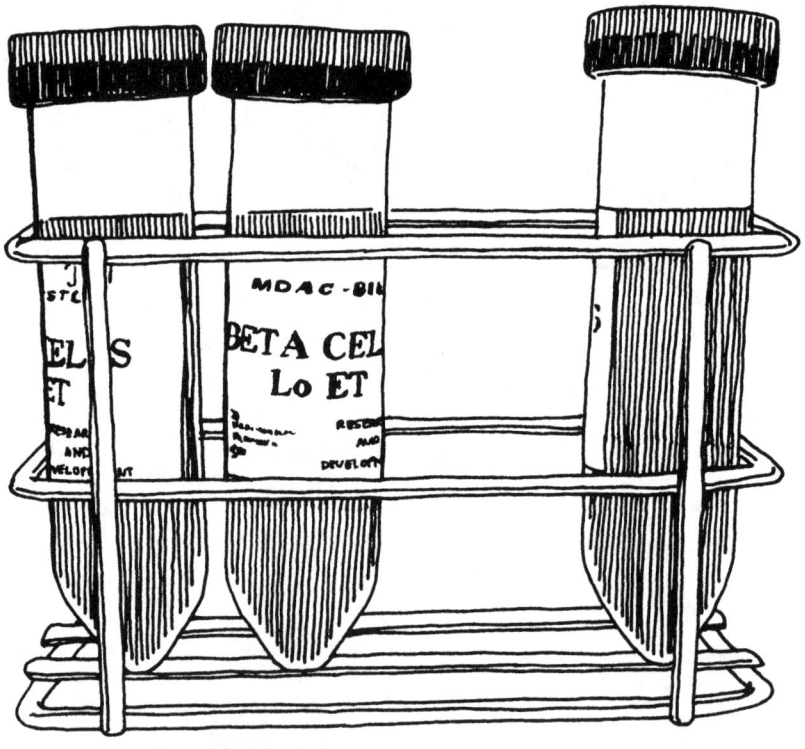

The production of large quantities of pure beta cells aboard the space station benefits 3 million diabetics. When injected into their livers, beta cells can cure diabetes, the third largest killer in the United States, behind cancer and heart disease.

A goal of our space station medical research is to develop 15 new medicines in a 10-year period. Without this manned station, it is estimated that we could only discover 3 in the same time period on unmanned free flyers. Millions of sick people around the world will benefit from this continuous presence in space.

Billy. The Lovelace Medical Center in Albuquerque is working with Mc-Donnell Douglas on the production of monoclonal antibodies in this unit."

"What are monoclonal antibodies?" Billy asked. "That's a hard one to pronounce."

"The monoclonal process fuses cancer cells with cells with antibodies that fight cancer. These hybrid cells multiply like cancer cells but kill off the cancer cells. Our work up here revolves around not only producing this antibody material in a pure state but also in helping isolate specific antibodies and then matching them up with the kinds of cancers and other diseases that they can seek out and destroy. Since there are over 120 different kinds of cancer, we have our work cut out for us.

"We think that biotechnology is revolutionizing the pharmaceutical business as much as the silicon chip did for the computer industry in the old days. Leroy Hood at Caltech has said that we've learned as much in the past 20 years about medicine as we did in the 2,000 years before. Just think what lies ahead of us. We already know that we can use living organisms in computer chips and increase their storing capacity a billion-fold.

"Now I'm going to shut up for a while and let Wayne cover electropho-resis for you. As you know, this is his area of expertise and the reason he's up here on this tour. Wayne, why don't you fill in Mary and Billy on this area of the module that you'll be working in."

"Okay, well, basically, every biological and chemical substance has an electrical charge to it, and when we apply an electrical charge to a mass of material in a buffer fluid the various elements pull apart according to their inherent charge characteristics and separate in EOS. That's what we call electrophoresis operations in space. A continuous stream of protein mate-rial is injected into a buffer solution flowing through a rectangular cham-ber. Here, you can see where the material comes in.

"Now when the charge goes through the buffer material, the various proteins pull apart and flow out of the chamber up here in separate streams. And since they are all separated, we can then collect those parts we want to keep. Due to the lack of convection and sedimentation in microgravity we can increase the voltage and get greatly improved purity over what we get on the ground. Band spreading doesn't occur up here and we can use bigger inlet ports on the station and quadruple the amount of material we process in the same amount of time. It's four times purer than the material we get on Earth, and this has great advantages in the long-term treatment of disease. We can also increase the sample concen-trations 100 times more than on Earth without getting sample stream collapse. The bottom line is that we can separate 700 times more material up here."

"What are you going to be doing at the EOS unit on your tour, Wayne?" Mary asked.

"I'm going to be calculating the flow rate in the separation column several times a day. I'll be doing that by injecting an air bubble into the sample pump. By measuring the bubble's rate of progress within this column right here, I can calibrate the pump's operation to meet our target production rate. In this case, it's erythropoietin. That's a hormone that your kidneys usually produce. It controls the red cell production in your body.

"And, since red blood cells only live four months or so, if your body quits making them or doesn't make enough of them, you end up being anemic. About 1 percent of the people in the United States suffer from this kidney disease. We've found that by coupling the technologies of EOS and

At work at the electrophoresis unit, Wayne Morrison is researching a method for stopping the uncontrolled growth of cancer cells, America's number-one killer.

genetic engineering, we can produce ultrapure erythropoietin at a much lower cost than we originally predicted. McDonnell Douglas got the ball rolling on this product.

"I'll also be doing some work on the islets of Langerhans project that McDonnell Douglas and the Washington University School of Medicine have been working on. The result is a new treatment for diabetes. What we're doing right now is fine-tuning our purification of the live insulin-producing islets of Langerhans from pancreatic tissue. The result will be islet transplantation, which can successfully control diabetes in most cases if we catch it in time. At least that's what our clinical tests have shown so far.

"I'm also going to be sending back some secret experiment samples on the space mail system for further study down on Earth while I'm up here on my tour. As you may know, this rapid transit system lets us deliver small samples of perishable and fragile materials to ground laboratories within 18 hours of the time we terminate an experiment. And several are due to be sent down during our stay. I'm responsible for their timely delivery. All I can tell you is that they may be useful in the treatment of cancer and are tied in to stimulating the aging in cancer cells. I think I'd better shut up now. I may not be asked back."

"Well," Edie said with a smile, "that was a pretty full morning. We covered a lot of ground. I don't know about you guys, but I'm ready for lunch. Let's see what we can find to eat."

"Sounds good. Let's go."

10

StarEagle

The next morning they all crowded into the commander's work station.

"Well, this is the big day, folks," Stu said. "You can all squeeze around here with me if you want to see the simulation on the screens. There are times when this place doesn't seem quite big enough."

Mary, Billy, and Wayne were eagerly anticipating what was to be a historic day aboard *Friendship:* the first trial run of the new laser fusion-powered spacecraft, code named *StarEagle*. The exotic system had been dreamed of for many decades. Way back in the 1950s there had been talk of propelling spaceships to high velocities using small nuclear bombs to provide the energy.

Then research in another area paid off when Arthur Schawlow and Charles Townes first proposed the idea of a laser in 1958. It took several decades of work, but the technology was finally developed, using lasers to initiate the small fusion explosions necessary to propel a spaceship. The process is called inertial confinement fusion (ICF) and the spaceship an IFR (inertial fusion rocket).

"Gather round, boys and girls, and watch Uncle Stu's magic picture show!" Stu said with a chuckle, as he fed commands into the on-board simulation system,

"Let me think out loud now as we go along. Stop me if you have any questions."

As he said that, an image of Earth appeared on the computer graphics terminal in front of him. It was a view from 50,000 miles above the North Pole; the continents were outlined with grid lines showing the latitudes and longitudes.

"That's too close to see what we'll be doing," Stu said to himself.

And, as he fed more commands into the system, the scene changed to one showing Earth and the moon. Suddenly, small pinpricks of light began to blink off and on in the vicinity of Earth.

"The blue dot is *Friendship,*" Stu pointed out. "The red blinking dot is *StarEagle.* It is now in a GEO parking orbit 22,000 miles above us and on the other side of Earth. Our test today is really a baby step to what we hope the system will develop into, but for now we're going to do it one small step at a time, just like NASA has always done. We're going to initiate our energy input and build up thrust gradually, one target pellet a second, to see if the old boat holds together all right. Today we're just going to loop the moon as sort of a dry run. Control is actually coming from our ground station and we're just eavesdropping on the test."

"Was *StarEagle* built here at *Friendship?*" Billy asked as the screen changed to a closeup profile of the waiting spaceship.

"We assembled the sections up here, but the actual construction of the payload module and the power stem was completed on the ground. They were brought up in sections on the shuttle. You can see here in the diagram that *StarEagle* looks a lot like a long arrow shaft. The payload is the arrowhead out here, and the shaft is where the laser does its thing with the fuel pellets, which dance along its length until they reach the nock end of the arrow where the feathers usually are. At that point the microexplosions take place and provide the thrust."

"Where are the pellets made?" Wayne asked.

"Okay, let's run through it one time," Stu answered, tapping the keyboard to activate another closeup view of the long *StarEagle* unit.

"This is the payload. For our purposes today, that end of the arrow contains monitoring systems as well as on-board cameras. We should end up with some fantastic views as the thing builds up speed and loops around the moon.

"Now, right behind the arrowhead are the fuel tanks. This is our pellet factory. The work we've been doing here on *Friendship* is being given its big test with the pellets being ingested into the system today. Basically, they're about the size of a BB, about half the size of a small pea. They contain the nuclear fuel, the deuterium enriched with tritium that allows us to make this small star explosion happen. We're making small stars out

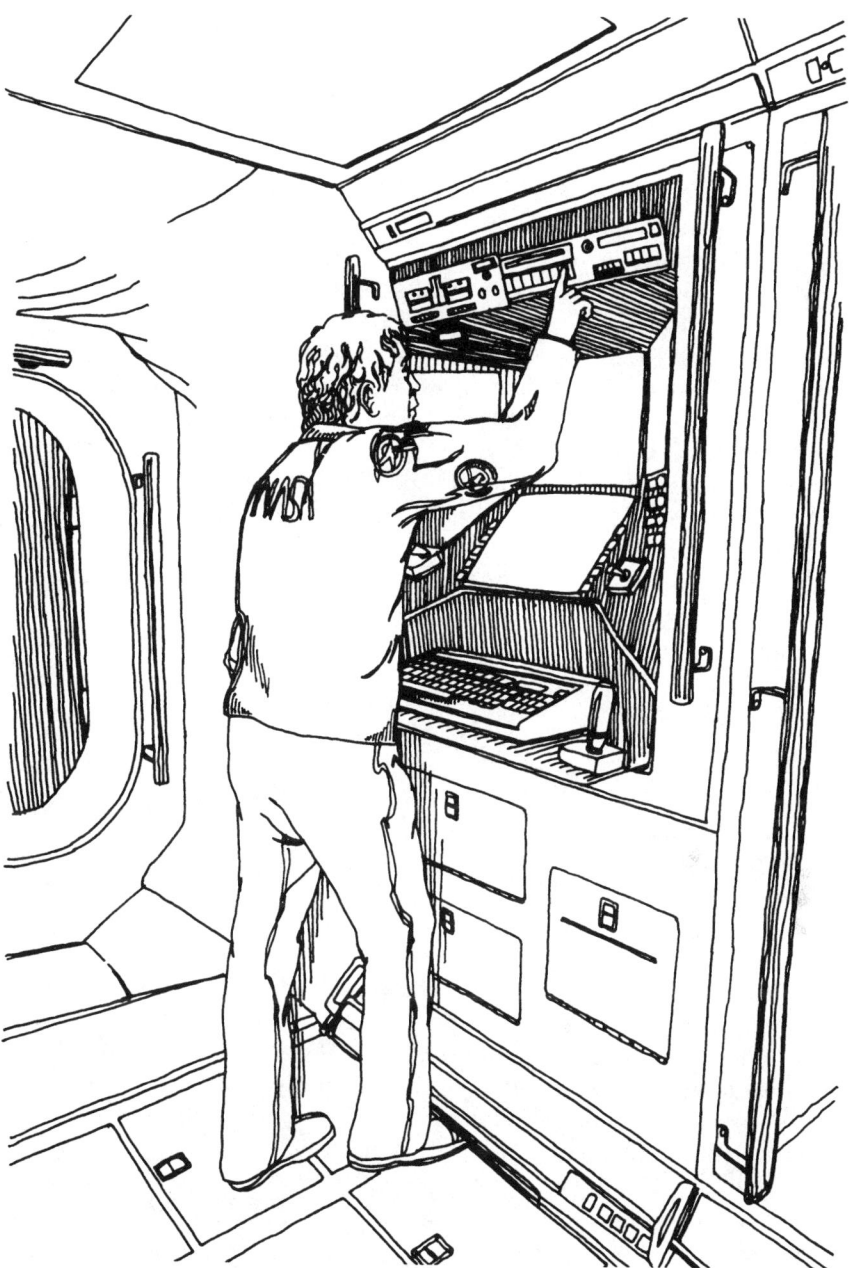

The space station commander on this tour, Stu Robinson, at one of the control center work stations. Note that he maintains his contact with the floor and his local-vertical by hooking his foot under the utility fixture.

of the pellets; we're creating small suns and we're directing the energy from these stars to drive our starship.

"Let me call up a cutaway view of the target pellets. Maybe that will help you understand. I think we've got one stored away somewhere in here. Yes, here it is."

A large cutaway drawing appeared on the screen. At this scale it looked like a basketball of many layers with a small golf ball in its interior.

"Okay," Stu went on, "you've got to remember that this thing is really the size of a BB. This cutaway should help you understand it better. The small ball in the center is the DT ignitor fuel, and the other layers are fuels and ablators. What happens is that the laser hits the pellet out at the end of the arrow shaft, and the pellet implodes. It just simply compresses in on itself to a density 20 times that of lead. And at the same time it heats to 100 million degrees Celsius; that's six times hotter than the center of the sun. This all happens in about a tenth of a billionth of a second, and the result is thermonuclear fusion.

"At the moment this occurs, inertia holds the BB material together. The deuterium and tritium nuclei fuse to form a heavier atom, and the result is energy released outward in an 'explosion.' We end up with about 200 times more energy than it took to make it all happen; in other words, we gain energy.

"This all happens so quickly and with such power that it's just stunning in its energy. As a matter of fact, the plasma that forms as a result expands inward with such power that the thrust created is about 100 times the thrust of the space shuttle launcher. We use magnet coils out here to direct the thrust out the back of the arrow."

"Let me see if I get this now," Mary interrupted. "There are three major parts to this system: the fusion pellet, the laser that ignites it, and the thrust chamber over here that directs the energy."

"Right you are. The laser is a KrF* gas excimer laser, and we're using geometric-stacking techniques in order to achieve the high peak power necessary for ICF capsule implosions. We get about a 9 percent efficiency."

"What's the advantage over the way we've been doing it all along?"

Stu laughed. "Oh, about the same difference as there is between a paper airplane and our suborbital *Orient Express.* Seriously though, when it's all perfected, this system will give us routine access to all of the planets in our solar system. Just to give you an idea, our first trip out to Mars took 210 days. This *StarEagle* system will take just 9 days once it's up and running. Now, how do we do it?"

"Since we're testing the IFR today, I would guess that that's the secret," Mary answered.

"Right you are. As we learn how to inject the fusion pellets sequentially at a faster and faster rate we will eventually hope to reach 10 to 20 percent the speed of light with the IFR, with *StarEagle*. At 10 percent the speed will translate to 18,600 miles per second. Of course we have to build up to it and then brake at the other end. In between we have a long coasting portion of the flight when we power down the system.

"But once we have our *StarEagle* fleet in operation we'll have routine access to our own backyard solar system and we'll be able to use its resources to the max. Jupiter, for example, will be just 39 days away in the VIP mode, 93 days in the cargo configuration.

"Eventually we'll have the equivalent of interstate highways between our neighboring planets. Then when we get real brave, we'll head on out into the stars and tackle interstellar travel. But by then this IFR technology will be pretty old-fashioned and we'll undoubtedly have something even faster."

"*Friendship*, Houston."

"Go ahead, Houston."

"We are about to initiate thrust; are you all up to tracking with us?"

"Confirm departure zone capability, Houston."

"Okay, Stu. Hang onto your hat. We've all got our fingers crossed down here. Sure hate to hear the howls in Congress if this sucker blows up."

"Oh, ye of little faith, Houston."

"Stand by."

Stu turned to the rookies. "We're testing a new ICF system today. The terrestrial system just wasn't suitable for use in space what with the high operating temperatures, so we're hoping that we have that part of the puzzle solved. Let's listen to Houston; here on the terminal we can watch the simulation of what's supposed to happen."

"Laser on."

"Pellet accelerator in standby."

"Capacitor bank confirmation."

"Radiator integrity check, mark."

"Mirror angle plus or minus parameters."

"Start up reactor monitor."

"Activate magnet coils."

"Fire pathfinder pellet."

"Confirm angle."

"Temperature readout nominal."

"Thrust chamber is go."

The group gathered around Stu and watched as *StarEagle* seemed to hesitate and then moved almost imperceptibly. The simulation seemed to be ahead of what they were hearing on the voice loop from Houston. Stu tapped his keyboard and the view shifted from a closeup of *StarEagle* to another polar view of Earth, with *Friendship* and *StarEagle* once more noted as blinking dots. Slowly, *StarEagle*'s red dot began to move as the velocity change built up. The fusion-powered arrow was now beginning to increase its Delta V and would continue to do so for this portion of the test.

"Confirm one pellet per minute rate."

"Someday," Stu said, "we expect to detonate 100 pellets per second in our power-building phase and then it will really become a screaming Eagle. But for right now we just hope to get it up to one per second. We're not really sure what this will do to the stresses on the stem. We think we know from our computer models, but you never know until you get on the horse."

"Continue one pellet per minute rate. Torque check?"

"Go!"

"Aft temperature?"

"Go!"

"Radiator flow?"

"Go!"

"Great downlink picture now. We'll put it on the network in a moment. Let's make sure this beast holds together first."

"We're going to take *StarEagle* out to the moon today," Stu said, "and let our free-return capability bring her on home. Our propulsion test will only last for another minute or so, but we expect to reach 100,000 miles per hour before we shut her down and do our braking test.

"When we switch ends and fly tail first we'll again use the fusion explosions, but this time we'll use the energy to slow us down just like we do when we deorbit burn the shuttle. Then we hope to get the free loop around the moon and head back toward Earth.

"We'll park *StarEagle* up at GEO for a couple of days while we check all the data carefully, and then we'll send our OTV up to get her and bring her down here so we can disassemble the stem and take parts of it down for analysis. And if you're wondering, Peter and Kathy are out of harm's way. They might be able to get some pictures of *StarEagle* if we're lucky. Their flight plan calls for them to be tracking all of this if they can."

The red blinking dot that was *StarEagle* was now moving across the screen in front of them. The arrow had been loosed at the target moon.

"Let's jump ahead a couple of days on this old machine," Stu offered, "and I'll show you our rendezvous simulation between the OTV and *Star-Eagle*. This system that we use on board for this kind of thing was developed by Don Eyles and his friends at The Charles Stark Draper Lab in

Cambridge. It uses Digital Equipment VAX computers and Evans and Sutherland graphics equipment. It's really tremendous for permitting us to practice darn near anything we want to before we have to perform the actual operations themselves. And we also use it in real time for data processing as events take place.

"It's like having eyes all over the place. We can change our point of view from inside *Friendship* here on the flight deck to outside on board the OTV as we approach *StarEagle* after we get her back. We can step outside ourselves and look down on Earth from 500,000 kilometers away if we need to.

"And we can change scale within the scene if we want to. Let's say we want to keep Earth and *Friendship* in their true size relationship but want to enlarge *StarEagle* and the OTV for our initial maneuvers. No problem; all we do is call up the proper displays and this joystick lets me practice docking with *StarEagle* before I actually have to do so in real time."

"Shutdown."

"Confirm, Houston."

"Set sail for coast mode."

"Delta V check."

"Midcourse not necessary."

"Way to go, gang; so far so good!"

"Good show, Houston!"

"Thanks, Stu, keep your eye on her."

"One other thing we can do with this simulation system," Stu said as his attention returned to the rookies, "is to go over *Friendship* inch by inch. We can set the point of view for off our starboard bow and watch *Friendship* as we bring *StarEagle* into port. Or, we can stand at any point on our local-vertical, local-horizontal. The entire space station is inside the computer, so we can plan out everything and every move we think we want to do with incoming spacecraft before we actually have to do it. And we can add things in, like where we will have work crews doing EVAs on the day we want to bring things in closer with the OTVs, and we can work around ourselves. One hell of a system."

"Let me get this straight," Wayne interrupted. "You have all of this preprogrammed into the computer. But what if you want to 'see' or track a new satellite or spacecraft that's not in the computer? What then?"

"Good question, Wayne. We have the capability of assembling just about any object we want to by using basic building blocks within the system. We can even add up to three moving parts to each spacecraft. The system uses what you might call generic parts consisting of flat panels and cylinders to construct just about any object we want to practice with.

Then, when we add the vectors and transformations that pertain to the spacecraft or the star fields and planets, we create a dynamic scene of what we want to see. And we can choose from four coordinate systems.

"There's the body system that gives us the point of view of the spacecraft, and the local system that moves with the spacecraft—that's the local-vertical, local-horizontal that I mentioned earlier. We also have an inertial system that is locked onto the stars and an Earth-fixed system that rotates with the planet below us. And as you can see, we can choose from a wide range of color graphics to keep everything separated in our minds as we go through the simulations. Really a very valuable tool to have on-board."

"*Friendship, Molly Brown.*"

Stu turned to the communications station area. "Read you five bye, *Molly Brown.* What's up?"

"Hey, Stu. Just wanted you people to know that we have locked on with *StarEagle* and are tracking her."

"Super, Kathy. That's better than we hoped for. We ought to get a real good look at turnaround and retrofire now. Think you can get any footage of it?"

"Say again, *Friendship.*"

"Think you can get any footage of retro?"

"We'll do our best, Stu, but until we get confirmation of *StarEagle's* apogee we won't know for sure. We do plan our own burn for transEarth injection so that we can caboose *StarEagle* back home. That's on our flight plan anyway. If we miss the retrofire, maybe we'll at least be able to get some good shots of the bird approaching home with Earth in the background. Suppose that'll make the evening news?"

"Sure hope so, *Molly Brown.* Good hunting."

"We're gone, *Friendship.*"

"Well, troops," Stu said to the rookies, "that's it for a while. We'll stand by and listen in to see how *StarEagle* is doing and whether or not Kathy and Peter can get a good look at her. Any more questions?"

Mary cleared her throat, but did not speak.

"Mary," Stu finally said, "you look perplexed."

"Well, I am. Is it realistic to expect to reach even 10 percent the speed of light? I mean, really. That's almost . . . no, it's *over* a million miles a minute."

"Very true, Mary. But you must remember that it's in a vacuum and it's a slow, steady buildup in velocity. You can't use your Earth frame of reference when you think about these things. I know 66 million miles an hour is unbelievable. But so was walking on the moon just a couple of generations ago. Or building a space station. You have to adopt an interstellar base of reference."

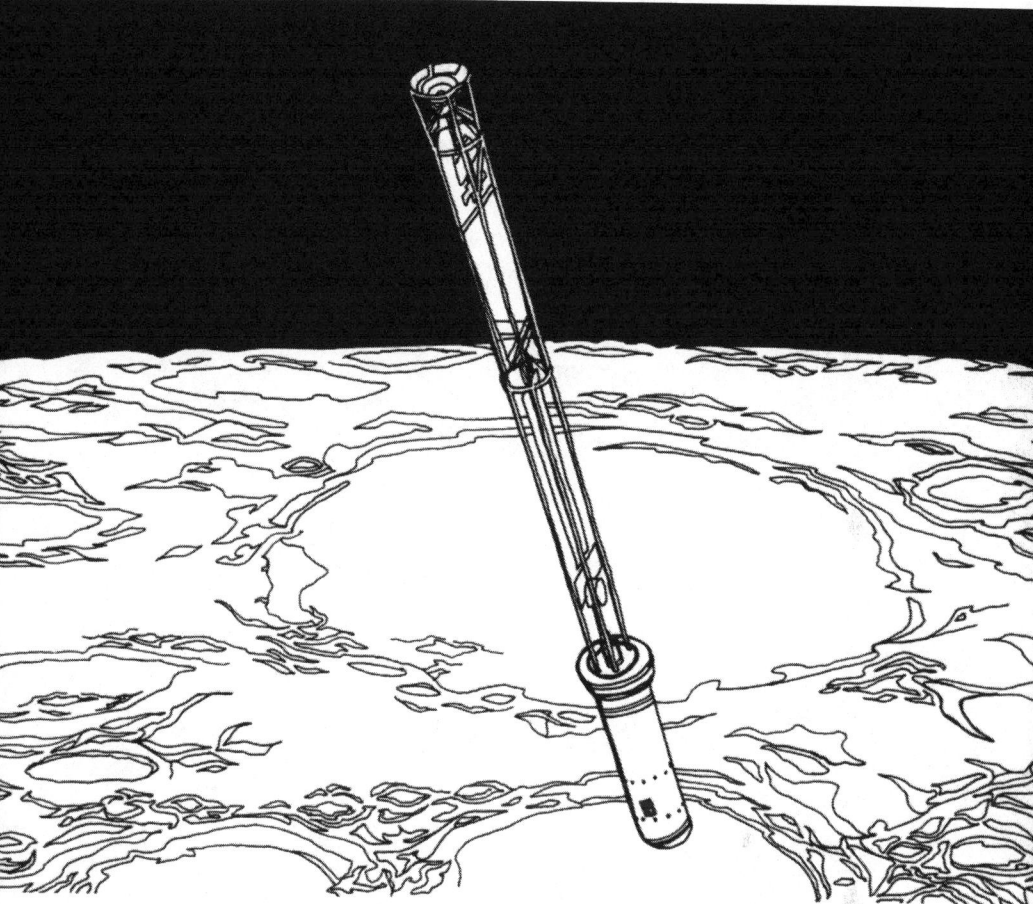

The test of the fusion-powered spacecraft, *StarEagle*. It will someday travel at 20 percent the speed of light and will greatly reduce interplanetary travel times. A boon to Earth's energy needs in the twenty-first century, fusion power will be based on the energy stored in ordinary seawater.

"I guess so, but it sure blows my mind."

"It does mine too, sometimes. You know, one other thing that you need to keep in mind with all of this talk about inertial confinement fusion is the tremendous impact it's going to have on our Earthside energy needs. At the rate we're going, we are simply going to run out of fossil fuels before the end of this twenty-first century. That means no more coal, oil, or natural gas. And, since we expect the world's energy demands to increase over seven times in the next 200 years alone you can see the pressure there is on us to develop alternate technologies. We're just about out of dead dinosaur juice. We need to come up with something else. This will also help us out on the acid rain problems and maybe alleviate the greenhouse effects we've been monitoring.

"Unlike conventional nuclear fission power plants, fusion plants are very safe, and they leave no radioactive wastes, although there's a small amount of radioactivity generated in the reaction chamber itself. All we need is the fuel to trigger the fusion reaction, and that will be almost free once we perfect our methods of mining it from seawater. There is as much energy available to us in just one gallon of seawater in a deuterium-tritium reaction as there is in 760 gallons of oil, and it costs about 5,000 times less than coal or oil in units of energy delivered.

"Well, so much for the sales job," Stu said, smiling. "Now, who wants to be first playing with the joystick here?"

11

The Curiosity Shop_____

"Look down there," David Bishop said, "bet you've never seen dawn at the Cape from this perspective."

It was just after breakfast for the blue shift aboard *Friendship,* but it was well into the dawn of an early autumn day below them along the Florida coast. The rookies were gathered in the observation node with David, the ship's astrophysics officer, who was filling them in on the work being done way out 'there.' Down below they could see their spent launch pad off to the north of their orbital path. It was an unusual sensation to realize that far below them papers were just now being delivered in the quiet neighborhoods opposite Pad 39B in Titusville.

"Who knows what we're trying to find out?" David asked them quietly as they looked out into the universe.

"You mean what are we doing up here in *Friendship* as far as the universe is concerned?" Wayne answered.

"Yes. What questions are driving our work in astrophysics, astronomy, and whatever?"

"Well," Mary said, "I suppose we're trying to find out how the universe is working, what it's made of, and how all of this can affect those of us on Earth."

"That's a good start," David smiled. "What else?"

"How big is the universe? And how old?" Wayne added.

"Those are all good answers, but there are more," David said. "And we have to be very careful that we do not create our own answers by the way we ask the questions. We want to know the truth, not what we want the truth to be, or not what we think the truth to be. Billy can back me up on that, since astronomy is his thing, too. He knows how basic some of our questions still are.

"Let me try to cover some of the areas we're looking at right now from *Friendship* as well as on the ground and by way of some of our orbiting tools that we have linkage with. Don't hesitate to ask questions as we go along. That's one thing we're not short of—questions.

"Most people have looked out at the stars and wondered what they are. For centuries we've had so little real understanding of them. It has only been fairly recently that we've learned that most of the starlight we're looking at left its source even before there were people to look up at the stars. What we're looking at out there in many cases is not even there any more, but its light is only now reaching our eyes. When we stand out in our backyard at night and look up at the stars, we're really seeing back into time the way it was millions of years ago. For all we know, we might even now be halfway over the threshold of another universe out there.

"Let's see, you people came up with most of the basic questions we're trying to ask. Just be careful that you don't make the false assumption that many people make, and that is that we already know most of the answers—and that all of the time and money we're spending up here is just a waste of time. I assure you, it is not.

"What we're really trying to do is to understand our existence and what our future is in it. And it gets pretty damn complicated. There are so many theories you couldn't believe it. We're trying to find out how the universe began, how it operates on a day-to-day basis, and what's going to happen to it.

"How big is the universe, and how old is it? What is in it? And what is the stuff of the eternal night made out of? How far away are the other galaxies, how big are they, and what do they have in them? Are they made of the same stuff that we are? And what is that stuff? We call these things out there stars and quasars and black holes. Those are just the names that we've given them. What does God call them and how does it all fit together and work?

"You are existing right now here in this observation node, but how is the stuff in you, in your minds and bodies, related to that star stuff out there? Is all of this really just existing in a single cell in the body of the

Supreme Everything, or are we all there is?"

"Wow," Wayne broke in, "this is heavy stuff."

"I know it is, Wayne, but we have to ask these questions. It's part of our nature as people made of star dust to ask these things, to be curious, and to be continually reaching out for other valleys, other shores, and other stars."

"You started to talk about theories," Wayne offered.

"Yes, theories. Most of them are so very difficult to understand that just a handful of people in the world can keep on top of them. Someone has a flash of insight, and the next thing you know we have a stunning new way of looking at the universe and everything in it.

"Einstein just had gravity and electromagnetism to work with, and he combined them into a unified field theory in 1929. Quantum physics looks at the microscopic world, and gravity gives us an understanding of the macroscopic world. Relativity theory concerns time, space, mass, motion, and gravity, too. And we've discovered two other basic forces of nature to go with electromagnetism and gravity—the weak force and the strong force. The strong force glues quarks together into protons and neutrons, and the weak force causes certain types of radioactive decay.

"Then, not too long ago, John Schwarz of Caltech and Michael Green of the University of London came up with a new theory of the universe that they called super-string theory. That's a theory of things where the fundamental building blocks of energy and matter are not just infinitesimal points but very tiny strings. Their universe has 10 dimensions. And another popular theory, known as supergravity, has 11 dimensions instead of the 4 that Einstein worked with.

"Einstein's world of electrons and the nuclei of atoms has grown a great deal smaller with all of this new research. Now we have the quarks that I mentioned, plus gravitons, gluons, neutrinos, Ws, Zs, bosons, and fermions. We have symmetry and super-symmetry, GUTS for grand unification theories, and even a TOE for the theory of everything. We've searched for WIMPS, or weakly interacting massive particles, and also for evidence of proton disintegration. We now know that atoms have a planetary structure. What we'd like to do is write one single formula that describes everything in nature, from the smallest to the largest forces and particles.

"Particle physicists look inward into the tiny universe within the atom and cosmologists look outward into the universe, and both are beginning to accept the fact that they are both looking at the same thing.

"Stephen Hawking and the others know that the universe is 14 to 15 billion years old, and they've traced its roots back to within a billion trillionth of one second after its creation, after what some people call the Big Bang. They're trying to discover if all four basic rules of nature that I mentioned before existed as one entity at the moment of the creation. They're trying to unify the four basic forces into one, but gravity is giving them fits. They call the moment when all of the matter in the universe was concentrated in one spot 'singularity.' And they now think that at the moment of the creation all of this matter exploded and created not only the universe but also space and time.

"They also believe that the reverse happens when a star dies and collapses inward on itself, creating a black hole. They say that space and time cease to exist in black holes and once more singularity happens. Hawking says that there may be a thousand million black holes just in our Milky Way galaxy."

"Wait a minute," Wayne interrupted. "Understanding the whole universe and how it works is a bit much right after breakfast. I can hardly keep up with all of this. A lot of it I've studied in my course work, but I'll have to admit some of it sounds like a cosmic fairy tale. And it seems to me that these guys gather a lot of moonbeams and make pretty big mountains out of molehills."

"That they do, Wayne, but we'll never be able to recognize the answers if we don't ask the questions from every conceivable angle, and that is just about what these theoretical guys do for groceries. It takes a lot of GUTS to keep from stubbing their TOES, as the old joke goes. But let me cover just one more aspect of the theories before I move on to the next subject.

"We're trying to discover if the universe is in chaos or if there is a smooth unity to it. And one of our newest tools is the superconducting super collider that we're using to gather information about high-energy physics. If you're not familiar with it, it is a ring of magnets over 120 miles in circumference. It's used to smash matter into even smaller subatomic debris than we had back in the eighties. It's another step beyond the CERN facility in Geneva where they discovered the W and Z particles in 1983.

"That accelerator produced an energy level of over 100 billion electron volts; that's over 1,000 trillion degrees, roughly the same temperature when the universe was 10^{-12} seconds old. At 10^{-20} seconds, even further back in time, we are at the moment when Hawking thinks black holes were first formed. And at 10^{-32} seconds the entire universe is about the size of a softball at a temperature of 10^{27} degrees Kelvin."

"The size of a softball?" Wayne asked incredulously. "You've got to be kidding. How could all of these planets and everything fit into something that small?"

"I know it's difficult to grasp, but take away space from your thinking and it's a bit easier to consider the possibility. Now, the fellows at CERN hit a wall when they got back to 10^{-43} seconds. Matter of fact, it's known as the Planck Wall, a moment in creation beyond which the scientists could not describe space, time, or matter as it existed. Don't ask me to explain the mathematics of it; it's way beyond me. But we are getting some clues from the super collider. It's 40 times more powerful than Fermilab's Tevatron and smashes protons and antiprotons together at 40 trillion electron volts. We're truly getting close to the answer."

"You keep mentioning this Hawking," Mary said. "Isn't he the fellow in England at Cambridge that has been living his life in a wheelchair?"

"Yes, he is. He has Lou Gehrig's disease, amyotrophic lateral sclerosis. He got it his first year in graduate school where he was studying theoretical physics. He was only given a couple of years to live, and at first he gave up and even started drinking pretty heavily. But for some reason he realized that what he was studying was a cerebral subject area, and he didn't really need his body, just his mind.

"Hawking's body deteriorated and he became restricted to a wheelchair. His speech could barely be understood by any but his closest family and associates. But he swept aside the disease from his mind, married, and had three children. *Time* magazine called Hawking an equal of Einstein, and his goal has been to understand nothing less than how the universe works.

"I've often wondered how many glial cells Hawking has in his brain. I remember reading somewhere once that after Einstein died they tracked down his brain after the autopsy and found it dissected and in pieces in glass jars in a box stored behind a beer cooler in Wichita, Kansas. The doctor who did the autopsy had taken it there. And a neuroanatomist from the University of California at Berkeley discovered that Einstein had 73 percent more glial cells than the average person. Glial cells are a part of the universe of our bodies.

"Well, enough of that, what else would you like to know about our work here in this area of *Friendship?*"

"I know there is a lot of equipment out there in orbit and also at the top of *Friendship* pointed out toward the stars," Mary said. "What are you looking at?"

David gestured toward the stars. "I started to get into that earlier and somehow got sidetracked. Basically, the universe and everything about it exists. We are not inventing anything—we are simply trying to define it. And we're trying to do it with an open mind, as I said before.

"You know, about 90 percent of the mass of most of the galaxies out there is invisible. All we see are the suns like ours, the ones that shine. We don't know how many of them have planets around them, and if they do, if they are anything like our Earth.

"On a clear night a person down below can see about 6,000 stars with the naked eye and nearly 100,000 with the help of binoculars. You probably have heard of our SETI program, the search for extraterrestrial intelligence, right?"

"Yes," Billy said, "we covered a unit on that."

"Okay. Well, in SETI we are searching the 773 stars within 80 light years of Earth for any evidence that anybody or anything is there. That covers a footprint of about 475 trillion miles. Now two-thirds of all the stars have a companion star. Our single sun is really in the minority as far as that's concerned, although there is a theory that every 28 million years a companion star to our sun comes back this way and disturbs things a great deal.

"An astronomer by the name of Piet Hut put forth what he called a death star theory that this is what has caused many of the extinctions on Earth in the past. They call this ghost star Nemesis, and one of our programs is trying to track alternative orbits for this star to see if we can find it and keep a handle on it. Richard Muller, an astrophysicist at the Lawrence Berkley Lab, is also involved in the search.

"But, getting back to my point. We're searching for another planet outside our solar system, in other words one that does not circle our small sun. So far we've found a cloud of dust circling Beta Pictoris about 50 light years up the road that seems to tell us that the raw material for building a planet is there.

"And at the 21-light-year mile marker there's something nearer to us traveling with a star called Van Biesbroeck 8. But that object is about 2,000 degrees Fahrenheit, and it's too hot to be called a planet. It's about the same size as our brother planet Jupiter, but with ten times the mass. We've given it the name VB8B. Both it and Jupiter could almost become stars themselves if they took a notion to start the pot boiling and brewed up some nuclear fusion.

"I mentioned the raw material. Stars form out of clouds of dust and gas that just sort of come together by virtue of gravity. We're studying these protostars, as we call the kid stars, to understand the various stages they go through in their development. Now with this star brew stuff, there is also the material to form planets. We're looking at several young stars right now on this tour to see if they have planets forming out of the matter we see in that area. Two of them are HL Tau and R Mon.

"I know you've probably heard of Barnard's Star; it's just out our back door at six light years away. It wobbles periodically, and that might mean that it has a couple of planets tugging on its gravitational field. They would probably be the size of Jupiter and Saturn based on the gravitational wiggle they're giving Barnard.

"Actually, there may be billions of planetary systems out there just waiting to be discovered and explored. Quite obviously, any migration out to them is a long way off, but by starting now we as a species hope to be able to escape to them by the time our sun begins to run out of star stuff."

"Are there any other possible targets for planet watching?" Wayne asked.

"Matter of fact, our old IRAS infrared astronomical satellite telescope studied 9,000 stars when it was in orbit back in the mid-eighties, and we think that 50 or so may have planets forming around them. Unlike the usual visible starlight, infrared rays can be seen through interstellar dust. Gerard Kuiper down at the University of Arizona really got us started in this field of astronomy back in 1965.

"Our IRAS satellite was so sensitive that it could detect the heat from a 20-watt light bulb on Pluto, 4 billion miles away from it. IRAS gave us 350 million bits of information every day for 300 days. In that time it observed 25,000 galaxies and over 250,000 other items of interest. It told us that there is about one new star every year being formed just in our own Milky Way galaxy. And it discovered a giant cloud of matter around the star Vega that might also contain planets as big as Jupiter.

"Then SIRFT came along back in the nineties—the shuttle infrared telescope facility. It was even more powerful than IRAS and let us confirm a lot of our earlier discoveries as well as add some more possibles to the shopping list. This infrared business is really pretty neat. It has given us quite a cosmic perspective, as Carl Sagan used to say."

"What other things are you looking for out there?" Wayne asked. "Not that I'm not already pretty impressed."

"Let's see if I can tick them off from memory. I already mentioned black holes, and we're studying the tremendous black hole at the center of our Milky Way galaxy. We didn't even know it was there until we got into the infrared technology 20-some years ago. It's right at the core of our galaxy, and it's producing one-tenth of our galaxy's energy, yet only takes up one-millionth of its volume."

"I'm curious," Mary interrupted, "which way is that from here?"

"Well, I don't see Sagittarius right now, but the next time you see that constellation it's just to the right of it, buried back behind the star clouds and interstellar dust. You can't see the center of our own galaxy with the naked eye—you need a radio telescope—but believe me, it's there. It's about 30,000 light years away from us.

"The center of our galaxy gives off about 30 million times more power than our sun. And it is about one light-day in size, about the size of our solar system, but incredibly more powerful. Did you know that not even light can escape from a black hole? Imagine what we could do with all of that energy if we could figure out a way to tap it. But I'll let Billy tell you about black holes later; that's his specialty.

"There's also a huge object on the edge of the constellation Virgo that we're trying to figure out. It is as big as 1,000 galaxies, and we think it is either a huge black hole or a cosmic string."

"You mean a string like in super-string theory?" Mary asked with a laugh.

"These are cosmic strings. And the theory goes that they're like seams that occurred during the time of the Big Bang. They are only as wide as a subatomic particle, but they exist through space for millions of light years, and the idea goes that they would attract matter.

"That's just one of the many theories of what makes up the 90 percent of the matter that we don't see out there. It may be matter—ordinary matter like we see all around us—or something more exotic like neutrinos, photinos, or axions. Whatever it is, we're trying to find out.

"Let's talk about quasars for a minute. So far we've found over 2,500 of them in the universe. They are about the size of stars but emit the energy of a billion stars. And they shine with the brightness of 100 galaxies. We're watching one of them known as PKS 2000-330 that's about 12 billion light years away. Most people think that they are the nuclei of galaxies and that they surround black holes that suck in the nearby stars and produce the tremendous energy we see.

"Matter of fact, the thing we see at the center of our galaxy might even be a small quasar. The big ones get really huge; one that we know of is 15,000 light years across and 600,000 light years long.

"Incidentally, we didn't even know that quasars existed until 1965 when radio astronomers discovered them. We really don't know what they are yet, but that's a good reason to be studying them from here on *Friend-ship*. Pulsars are the remains of old stars and some people think that quasars just might be the early stages of the development of new galaxies. And some people think that they're the remains of the original fireball explosion that created the universe. Some of them are speeding away from us at almost the speed of light, and they're the farthest things we see from Earth, out on the edge of the known universe."

"Dr. Bishop—" Mary broke in.

"Please, Mary, just call me Dave."

"Well, okay. But how can you tell these things are going away from us so quickly?"

"By the color in the individual color spectrums. Things traveling away from us have their spectral lines red-shifted; if they're heading toward us they display a blue shift. We call it the Doppler effect. Christian Doppler discovered the thing in 1842; he was an Austrian physicist. And then in

1913 an American by the name of Slipher found out that the distant galaxies are red-shifted. So, if a star or galaxy is moving away from you its wavelengths will be longer and shifted to the red.

"Another interesting way we're working with the quasars, the quasi-stellar objects, is in the area of geodetic astronomy. And in that area we're using our very-long-baseline interferometry techniques. To do that we have radio telescopes set up at different places on Earth. For example, there is one at Fort Davis, Texas, and another at Westford, Massachusetts. There's also one at Onsala, Sweden, and another at Wettzell, West Germany.

"What we do is measure the radio source coming from the quasars and then put all of the signals together into a mosaic to see what changes in distance we're getting in the movement of the Earth's crustal plates. We can also tell that days vary in length; they're not all exactly 24 hours long. And we can keep track of the Earth's wobble as it spins.

"I should also mention the work we're doing to support the gravity-wave astronomy experiments. These are all based on Dr. Einstein's theory. We're looking for gravitational radiation waves. We discovered some back in the sixties, thanks to the work of Joseph Weber at the University of Maryland, who has turned out to be the father of gravitational wave astronomy.

"It's another one of our new disciplines, this looking for ripples on the mill pond of eternity. When a supernova explodes it sends out waves of gravity in all directions at the speed of light. Collapsing stars usually send out waves that are 200 miles from peak to peak, whereas a massive black hole might send them out millions of miles in length. As I said, Einstein first predicted them, and it's part of his contention that gravity is really a curvature in space-time.

"We have three satellites orbiting the sun right now in formation. They're linked together by laser signals, and any waves that pass by them will register and be downlinked to us here on *Friendship*. The idea was developed out at the Astrophysics Laboratory in Boulder. Just think of the waves we'd get if two black holes would collide or if two stars would run into one another. Matter of fact, we have our eye on a couple of binary neutron stars right now."

"Can I ask you about the Hubble space telescope?" Mary asked.

"Of course, what would you like to know about it?"

"I've read about all of the amazing things it has spotted out there, but I really know very little about it. How long has it been up here in orbit now?"

"Let's see. I think *Atlantis* brought it up in 1989 or so. They used the *Atlantis* because it was the lightest orbiter in the fleet and the Hubble

The Hubble space telescope sees 14 billion years into the past. As big as a boxcar, it weighed 25,500 pounds down on Earth. Most scientists consider it to be the most important achievement in astronomy of the past 400 years.

weighed 25,500 pounds on Earth. The 94½-inch primary mirror alone weighed 2,000 pounds."

"How much did it cost when it was first put up?" Wayne asked.

"Can't remember for sure, but I think it was a little over $1 billion. But don't forget, most scientists consider it to be the most important thing to happen in astronomy in the past 400 years since the first telescope was invented. The Hubble, or HST, as we call it for short, is in a 310-mile orbit above us, but we service it from here. About every two or three years we have to replace its batteries, and we change out the solar panels every five years.

"It's really something. It can see 14 billion light years back in time; we figure that that's about 97 percent of the history of the universe. We can see stars 50 times dimmer than those we saw from our observatories on the ground, and we can see 350 times more of the universe. The Hubble can detect a star 10 billion times fainter than one visible to the naked eye. It could spot a firefly over 10,000 miles away. Its guidance sensors keep it on a precise target for up to 24 hours at a time so that it can gather enough light to study some really far-out things.

"The Hubble is as big as a boxcar, 43 feet long and 14 feet in diameter. And when they polished its mirror at Perkin-Elmer in Connecticut they used a computer. They actually measured deviations by interferometry and then used an epicyclic path to polish the thing. They used a spiral path starting at the center of the mirror and working out to the edge and then back with no turnarounds. This way they got fantastic depth controls."

"Does it have anything on it except the big mirrors?" Mary broke in. "I don't know very much about observatories."

"Yes, it does. It takes the light that hits the two mirrors and diverts it to five different instruments on board. There's a high-resolution spectrograph that studies ultraviolet light and can tell a lot about the physical and chemical makeup of celestial objects. Then there is a wide field and planetary camera to look at large areas of the sky. A faint-object spectrograph measures the spectral and polarization of the light from stars and nebulae in our Milky Way galaxy and in neighboring galaxies.

"We have a high-speed photometer to measure brightness and time variations of ultraviolet and visible light from our target objects. Finally, there's a faint-object camera built by our friends in the European space agency to photograph stars seven times farther away than ever before. Remember, the light we're looking at left those objects 14 billion years ago. That means it started on its journey to our eyes almost 9 billion years before there even was an Earth. Just think of that."

"Do we have anything even close to the Hubble telescope on Earth?" Mary Two Hawks asked.

"Matter of fact, it's odd that you asked. I just talked to them yesterday about something we spotted with the HST that Goddard told us about. Sometimes when the Hubble spots something of great interest we notify the people at the William Keck telescope on top of Mauna Kea in Hawaii. That's not far from where El Onizuka, who was lost on the *Challenger*, grew up. The Keck has a ten-meter mirror; that's over four times bigger than on the Hubble.

"It's the largest optical telescope ever made. It was built in the late eighties at a cost of $85 million. Mr. Keck donated the money, and it was the single largest amount ever given up to then for a scientific project. Caltech and the University of California built the thing. It has 36 hexagonal mirrors, each 6 feet across, and it's twice the size of any telescope up to that time on Earth.

"We work in conjunction with Mauna Kea. They do a great deal of follow-up work on what we initially find of interest with the HST. That frees us up to continue our search with the Hubble."

"I'm embarrassed to say that I don't remember who this Hubble was," Wayne said sheepishly.

"No problem. It's a question I get asked a lot. He did live a long time before you were around, Wayne. Edwin Hubble was one of our most famous American astronomers of the twentieth century. In 1924 he succeeded in observing individual stars in our next-door neighbor, the Andromeda nebula, and proved that there were galaxies out of our solar system.

"Then he and his assistant, Milton Humason, discovered that the universe is expanding, and this gave us the evidence we needed that the Big Bang theory was indeed possible. The two of them showed that galaxies were traveling away from us and the more remote the galaxy, the faster it moves. Matter of fact, we now call that Hubble's Law, and the coefficient relating the distance and the velocity of the galaxy is called the Hubble Constant. So naming our first large space telescope after Dr. Hubble was very appropriate."

"Is that about it as far as hardware is concerned?" Wayne asked.

"Have I mentioned COSMIC?"

"No," Billy answered, "I don't think you've covered that one."

"Okay. Well, COSMIC stands for coherent optical system of modular imaging collectors. It's one of our tools for searching for Earth-type planets out to a reach of about 30 light years. Another device that the people at the University of Pittsburgh's Allegheny Observatory are using is a multi-channel astrometric photometer. Most of us just call it MAP. George Gatewood at Pitt developed it. It can nail down a star to within a millionth of a degree of arc. MAP uses fiberoptics rather than the old-style photographic plates.

"Speaking of that, most people are using CCDs now in that kind of work. These charge-coupled devices employ a silicon chip that converts the photons of light into an electronic signal. They are 100 times more efficient at gathering light than the old photographic plates are. We're not just getting older, we're getting smarter. The jury is still out on whether or not we're getting wiser. Can you imagine what those people out around Barnard's Star will think about us when our radio and television programming of the past 80 years starts arriving there? Well, what else would you like to know?"

"Do you have any connection with the gamma ray imaging telescope that they made out of the shuttle's main fuel tank?"

"Yes, Mary. You can't see it from here because it stands off our starboard beam quite a ways, but we do provide periodic maintenance for that unit. It has the telescope in the aft section, what used to be a 96-foot-long liquid hydrogen tank. The whole main tank is 154 feet long, but we just use the back end. It was carried on up to orbit near us a number of years ago rather than being permitted to burn up in the Indian Ocean. We salvage most of them that way nowadays. The holding yard is off over in that direction. We have 25 or 30 of the tanks out there now.

"David Koch of the Smithsonian Observatory came up with the idea for the telescope. Martin Marietta Michoud Aerospace actually builds the tanks. We're studying things like matter–antimatter annihilation, nuclear interactions of energetic nuclei, and electromagnetic processes. It's a fine way of gathering scientific data with hardware that would otherwise go to waste, and it helped our relations with the Congressional budget committee a great deal.

"Did I mention the very large array radio telescope yet, the VLA? No? That's a very unique and ghostly group of antennas just an hour or so away from Socorro, New Mexico. There are 27 of them out there on an ancient lake bed in the Plains of San Augustin. They're 82 feet in diameter, and each of them weighs 212 tons. When all 27 of them are focused on one object in the heavens, they become in effect a single instrument over 21 miles in diameter.

"Radio astronomy is only 50 or 60 years old, but some of the radio galaxies that we're studying with the VLA are huge. Some span over 100 million trillion miles. They give off tremendous amounts of energy and we're trying to figure out how. It'll be nifty when we finally get a handle on it.

"Oh, I forgot to mention one other thing about Hubble. It does not make X-ray observations, so we have something called the AXAF also hanging around the area that we service from *Friendship*. That's the advanced X ray astrophysics facility. And, as I've already mentioned, we need to study high-energy radiation. It gives us very important data on black holes, pulsars, quasars, neutron stars, and other invisible things out there. We call some of them 'bursters' because of the way they spit out X-rays.

"The AXAF has Wolter Type I mirrors. One of our primary goals with the unit is to study stellar corona composition. There has actually been an X-ray Astronomy Institute established now because of all the important finds the AXAF has located. It can see X-ray sources 50 times fainter than anything we had before.

"Well, I'm sure I forgot a thing or two, but that should pretty well cover my end of the station. Any questions?"

"You've completely overwhelmed me—us—with information. It's a lot to think about," Wayne said. "But, since the three of us are new to this business up here and hope to make a career of it, what can you tell us about what's coming along in the next couple of years . . . let's say the next 20 years, out to 2030 or so."

"I assume you mean in the way of equipment?"

"Right, the hardware. Where is the current technology today in 2007 leading us?"

"Well, we've already got a leg up on you on that one. I'm really never sure how much I'm supposed to disclose on these things, but we have a large space telescope array on the manifests for the next couple of years right now. It will be in low Earth orbit with us and will include a number of 25-foot-wide telescopes covering the infrared, visible, and ultraviolet spectrums. It will be 100 times more sensitive than the Hubble, if you can believe that. There's a joke going around the ground crews that we might even be able to see back *before* the Big Bang with it.

"We also have a group of 100-foot-wide space-based radio telescopes on the drawing boards to extend the reach of the very long baseline array that I mentioned earlier. They'll be placed in position as far out as 600,000 miles and will really give us a good view of the black hole at the center of our galaxy.

"Let's see, what else? Oh, we have a long baseline optical space interferometer in the works to help us in our planet search. The first two telescopes in this system will be placed about 300 miles apart for the resolution we need to look for other Earth-sized planets.

"Then there's a high-sensitivity X-ray astrophysics facility and a more sophisticated gamma-ray observatory. All of these things that I'm mentioning are reaching further into tomorrow for us; they're all based on new technology that we couldn't even dream of when the space station was first put up here. But the National Commission on Space back in the eighties helped set these goals for us, and great observatories was one of their more important priorities.

"Here on *Friendship* we'll be responsible for the assembly and support facilities for most of these new projects and observatories. And there is surely no lack of targets for us. We figure right now today that we have over 100 billion candidate stars to look at. Just think what it will be like to receive the first definite message from another star's civilization. And how about our first view of our own galaxy as we look back toward Earth, lost in the sprinkling of stars?"

"Thank you, Dr. Bishop," Mary said, "you've really opened my eyes today. I'll never look at the universe the same way again."

12

Banana Strings and Long Walks_____

Next morning the rookies were in the small Ames Library area of *Friendship* with Sean Finnegan. The library was a small island of privacy that was often used for discussions and briefings. The dominant feature of the cubicle was its large viewing port, and the rookies could not seem to get enough of looking at the passing scenes down on Earth.

"I'm your MOM," the white-haired old man began with a chuckle. "Officially, whoever has my job on board during a tour is the manager of orbital medicine, but everybody calls us MOM, and so be it. My name is Sean Finnegan. I'll have your head if you call me Dr. Finnegan. Just call me Finnegan; everybody else does. Agreed?"

All the rookies smiled in response.

"I do double duty by being the corpsman and working in the life sciences area. Today I want to go over some of the psychological and physiological parts of the flight that you should be aware of right from the start. Some of what I'm going to cover will duplicate what you were told during your training. But we think it's good to repeat some of it now that you're up here and have a different perspective on the place."

"To be honest with you," Mary said, "they crammed so much in us that I have trouble remembering a lot of it."

"Well, then, let's talk about banana strings and long walks. Those, believe it or not, are some of the things that many people miss the most as a tour goes on."

"What do you mean, banana strings?" Wayne asked.

"People seem to have the damnedest things that end up bugging them up here. Not being able to go for long invigorating walks in the fresh air is one of them. And, as far as banana strings are concerned, little things like that are what some people really miss up here."

"I don't understand," Billy broke in.

"Well, a lot of people have little routines to their lives that they never even realize until they are taken away from them. Do you brush your teeth first or comb your hair first thing in the morning? How about breakfast? Do you wake up slowly or hit the floor running? For a lot of people peeling a banana in the morning and then pulling off the little strings is one of the things that they linger over while they're waking up.

"It's crazy, I know, but we've had people turn sour on board over just such little things as that. In the old days, you could put up with just about anything on board just for the thrill and satisfaction of it all. It was more like camping out. Now, with long missions like we're flying here, isolation, confinement, and risk can take their toll if you don't guard against them. And fresh bananas are one thing we can't bring up with us. They just don't last long enough before they turn black. Screwy, isn't it?

"Now, you're going to be here for 30 days, until the next shuttle comes up. While that isn't as long as most of us spend on a tour, it will be long enough for you to go through some of the same stages that people here usually go through. So maybe I'd better cover them with you at this point.

"The first week on board is usually one of elation. You've just realized a life-long dream of flying in space. In your case, you've broken new ground by becoming the first doctoral candidates in orbit as part of your thesis work. So it's normal to be flying high, as it were. This wears off as you get down to your work on board, and about a quarter of the way along a mission we begin to see a lot of sick-call visits and wispy symptoms.

"Then, about halfway along we really see people hit a wall of depression. In many cases their experience here doesn't live up to what they've been expecting. Being away from their normal daily roles also causes some people a lot of psychic pain. We all fill a lot of roles during the day. We are husbands, fathers, wives, mothers, employees, bosses, customers, and on and on.

"Most of us have really been something special on the ground. A lot of us have been high achievers. Some of us have Phi Beta Kappa keys and doctorates. People on the ground usually defer to us, or at least listen to our opinions. Up here nearly all of us are bright, quick, and at the head of our class. Some very serious identity crises have happened on board *Friendship*.

"You must realize that while you're here you will be pretty much withdrawn from your normal social matrix of friends, family, church groups, and society in general. This is a microsociety in a miniworld up here. When things start to pile up on you, and they will sooner than you might expect, then I want you to come and tell me about it. It doesn't have to be a big deal, just let me know you want to talk, and we'll come in here and shut the partition and talk it out. Agreed?"

The rookies nodded.

"I mentioned isolation and confinement as being two of the major things we all have to face up here. A couple of the others are the constant risks, the crowding, the deprivation of things that are a normal part of your life, and even the environmental sameness of life on board. Don't be surprised if you get into the depression I mentioned, or if you experience strange fatigue, decreased alertness, and an erosion of your motivation. You might even start to 'cocoon,' as we call it, and find yourself withdrawing from us as a group. Or you might withdraw from certain people in the crew because of conflicts or the way they comb their hair.

"This can all lead to a severe decrease in your productivity and even affect your physical and mental health. Don't let it get to you. There is absolutely no reason you should not come to me about it. It will not affect your future flights or your personnel record. I promise you that. As a matter of fact, the idea that you recognize these problems and are mature enough to discuss them is a real plus on your record."

"You make it sound as if you expect us to have problems while we're here, Dr. Finnegan," Billy said.

"Just Finnegan, remember? Well, as a matter of fact, a lot of people do have problems in varying degrees. For some it's what we call the 'polar bear eye.' That's an old expression used at our Antarctic station. That's what most people call insomnia. And while we're on the subject, let me talk about sleep for a few moments. Actually, I've got a whole list of things that I'm supposed to go over with you.

"Since performance and happiness are so tied into sleep, it really got a high priority when they built *Friendship*. That's why we all have individual cubicles that we can close off from the noise and light. But we also have other ways of dealing with the problem.

"First of all, we try to maintain a regular sleep schedule. We really don't care when you go to sleep, but you must be up for morning muster with the other people on your shift or you go on report. The first time it happens we'll usually overlook it, but it had better never happen again. You all have alarm buzzers in your rooms and we all try to look out for the other guy on our shifts. We encourage you to follow the sleep hygiene program when you get ready to hit the sack at night.

"As you may have been told, we dim the lights in certain areas on board to match what is actually happening at Mission Control on the ground. When it's nighttime in Houston, it's nighttime on *Friendship*. Of course, that gets a little bit ludicrous because of all the sunrises and sunsets that we see every day up here. But we try to provide as many Zeitgebers as possible to help you out."

"Say what?" Wayne asked.

"Zeitgebers. Those are the cues you get every day that you may not even be aware of that keep your biological clock running along properly. Things like sunrise and sunset. Those things that keep you purring along— your circadian rhythm, that's your body's rhythm during a regular 24-hour period. Some people up here tend to get out of sync with themselves and wonder why they get colds and are so irritable with everyone at such a supreme moment in their lives. Headaches, upset stomachs, and constipation all go along with it sometimes, too.

"Dr. Charles Winget of NASA Ames studied this in great detail back in the early space shuttle days. He also found that disturbing your body's natural 24-hour rhythm could affect any medication you may take while you're up here. Matter of fact, it opened up an entirely new field of biology called chronopharmacology.

"Winget found that the effects of medication vary depending upon the time of day that you take it. An oral antihistamine taken at seven in the morning will last 15 to 17 hours; taken at seven in the evening it will last only 6 to 8 hours. The heart medication digitalis is twice as potent taken at night. This means that a lot of very powerful cancer drugs, for example, will have fewer side effects if the dosages are cut down in tandem with proper timing in the 24-hour body cycle.

"A lot of people increase their chances for colds, headaches, viruses, infections, and intestinal problems when their circadian rhythms are screwed up. The various functions in your body all follow a rhythm, too. Pulse rate, blood pressure, metabolism, hormone secretions, heart rate, temperature, and kidney function all have their own patterns. Most people perform best when their body temperature is the highest, usually about three in the afternoon.

"A change in your circadian rhythms can also trigger emotional and psychiatric problems. A lot of people fall into a deep depression, have a feeling of hopelessness, and a loss of self-esteem about 72 hours after a shift in circadian patterns. It takes up to two weeks for your temperature and brain waves to return to normal. Incidentally, that's a good reason not to change work schedules any more often than every three weeks

"Everybody who has ever suffered jet lag is familiar with some of these problems. Now you three might already be suffering from the changes in your rhythms the past couple of days. You'd better try to do your most important work as soon as possible after you wake up for a while. And eat your biggest meal then, too. But be sure to cut down on the calories until your body adjusts; don't overload with food.

"Cut down on salt and coffee, too—although a cup of coffee when your body's temperature cycle is at its highest can actually help improve your performance. Take it easy on the beer allotment for a few days. They all can affect you when you're Earthside, too, and you should also cut back on the wine and heavy stuff when your rhythm is upset down below.

"No new medications until the body adjusts; use only your regular ones. And stay away from sleeping pills. They disturb your rapid eye movement while you're asleep, and dreaming is a very important part of your psychological well-being."

"Gee," Billy said with a laugh, "maybe that's why I've been so spacey the last couple of days. At least now I have an excuse."

"Yep, could be," Finnegan said. "Now, sorry if this seems gross, but you all must stick to the rule and change your underwear every day and your outer garments twice a week. That's why we have a washer and dryer on board. None of us wants to live with foul odors and soiled appearances. Saturdays and Wednesdays are the two days we change pants and shirts, so stick to the schedule. You'll find that it has psychological benefits as well as hygienic.

"All your clothes have been prewashed to reduce the lint, so have your towels. They do that on the ground before each flight. Lint fouls up our ventilation system and floats around all over the place. It can contaminate our food and equipment, and it causes us problems when we inhale it.

"You'll find that mealtimes are an important part of our lives up here. Food becomes a substitute for what we all miss in our lives—our loved ones and friends. You'll find that you tend to linger over meals a lot more, but don't worry about it. It's great for communication between all of us, and it cuts down on the problems we can sometimes get when subgroups form in closed societies like this one.

"You may have a tendency to overeat, but that's normal, and so is complaining about the food. It's no different than college. We have an open locker for ice cream, snacks, and soft drinks, so help yourself at any time. Just try to be considerate and don't overdo it. We're not too strict about lunch and breakfast, but we do encourage you to eat together as a group for the evening meal. Saturday night is steak night, just like it is on nuclear subs.

"We also celebrate birthdays and holidays, and we have mid-mission parties. It all tends to keep up our morale and is a good way to keep the mission fresh. Did I mention our new art acquisition?"

"No, I don't think so."

"Well, you may not know it, but you brought it up with you when you came. We're going to have the unveiling tonight after supper while we're all up at shift change.

"You probably know that you can decorate your sleep area just about any way you'd like. We don't even care if you hang the Playmate or Playguy of the month up there, just so it doesn't offend anyone and lead to friction. But in our so-called public areas we've always had a rule against any personal decorations for the harmony of all on board.

"We do, however, get to change our two art display areas whenever a shuttle drops in from the Cape. Those who care to can vote on what they want to hang here in the wardroom area.

"On your visit you brought up a print of John Constable's 'Weymouth Bay' that he painted in 1807. That painting gives us a comfortable view of ground level Earth, looking out over rolling hills. We miss that very much up here. We selected it and the other one from our art file in the videotape library."

"What's the second one?"

"Ah, that's my favorite of the two. It's Renoir's 'Picking Flowers.' It gives you the feeling of gentle summer afternoons in the garden. That's why we have our planters in this module, too. Did you know that you can have one for your own? Remind me after a while and I'll show you our selection of seeds. Now, let's see, where was I?

"Let's stay with colors for a moment or two. I'm sure you've already noticed our color themes, but just in case you haven't, we use blues, greens, and violets in sleep and recreation areas. They are supposed to be cool and restful and have a tranquilizing effect. Then, in our work modules, we have reds, oranges, and yellows. They're warm colors and are supposed to stimulate you. Both color groups have contrasting accents and trims. You also may have noticed that our lighting is diffuse, indirect, and nonglaring.

"You know about the personal communicators; I see you're all wearing yours. Our back-up system for them is the audio announcement system in all of the modules. That's also for general announcements and group paging, but for your day-to-day communications you should use your personal comm unit. You can post notices on the bulletin board here in this module or on any of the CRTs.

"As far as outside communications are concerned, you can use your own private line in your sleep area to talk to your families down below, and we encourage that once or twice a week. But I want you to be aware of a couple of risks of doing that. If by any chance you get bad news from home, say a death or serious injury, it's going to be very tough on you. Unless it happens in the next day or two while Smokey is still here with the orbiter, you're going to be stuck up here for the duration of your 30 days.

"The second thing shouldn't affect you rookies too much, but it does bother some of the older married people who fly. And that is an obsession that some of them seem to develop with calls from home. They get down-right paranoid about how their husband or wife might be behaving themselves while they're gone, or why they haven't called lately. It can lead to some very dusty mental health."

A two-way closed circuit television between *Friendship* and the ground was used for communication whenever possible. Communications were enhanced by the visual image, which was particularly important to those who were working with scientific teams on the ground. Routine communication was done by teletype, and there was really no limit on how much could be sent; they had plenty of capacity. Often writing home every day was what kept the astronauts on an even keel.

Regular television programming came from the commercial networks. Even soap operas could be taped on the sleep-area VCRs and then played back during off hours. The evening news broadcasts were rotated among the networks in the interest of peace and tranquility. Those on the day cycle with Earth taped the news broadcasts for those on the nightside shift with Houston.

"As long as we're on recreational time," Finnegan continued, "let me cover some of those items. Each of you has been provided with a personal Walkman so that you can listen to tapes as you work or during your off-hours. You brought along the kinds of music you prefer, and your favorite books.

"HAL also has a pretty complete library if you run out of reading material, although a lot of people don't feel comfortable reading off a screen. There is a very well stocked library of tapes, movies, entertainment specials, and sporting events. You can watch them in your sleep area, but we highly recommend group watching whenever possible for your own mental health. Withdrawing just isn't a very good coping mechanism. Naturally, if you're touchy on a given day and just want to be left alone we'll understand; just so it doesn't become a habit.

"We recommend that you force yourself to spend at least one leisure hour before you go to bed for the night. It will help you relax and help you keep things in perspective.

"Looking out the window has always been at the top of the leisuretime favorite list of things to do, and that's why we have so many of them built

Sean Finnegan burrows down into his nest in his personal compartment to enter notes in his journal. Such a diary is encouraged aboard the space station.

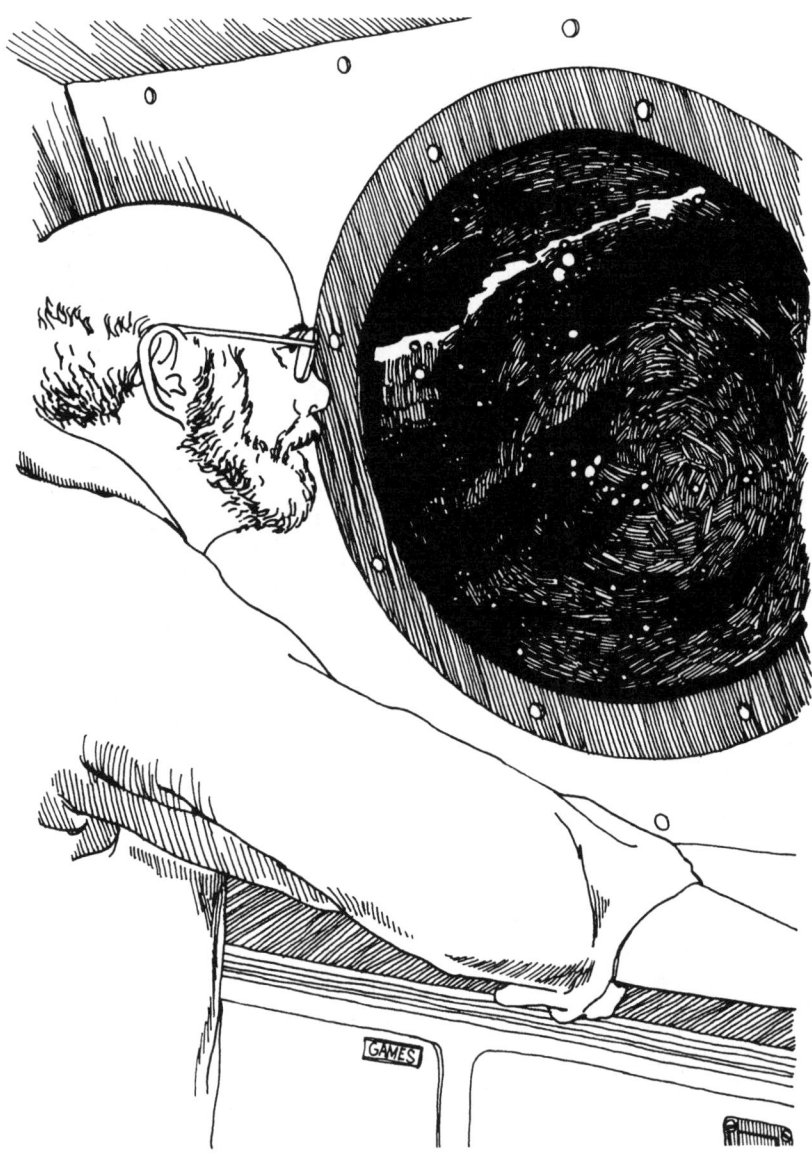

The ship's MOM looks out at the universe. Sean Finnegan is the manager of orbital medi-cine. Windows play an important part in the general mental health of all on board *Friend-ship* for extended tours. They also help them maintain distance vision, one of the serious problems facing submariners, who journey in similar-shaped vessels.

into *Friendship* and why we have the observation nodes. The windows can give you minivacations during the day; just don't get trapped into too much gawking, or your primary assignments will suffer and the ground will get on you.

"That's another thing that you should be aware of, the part the ground plays in your mental health. When we come on board *Friendship* we all want to get along for the duration of our flight. All of us hope to be selected to come back, or to grow in our careers as a result of our flight. Yet we soon find ourselves confined with people who may have irritating mannerisms and annoying habits. The guy who always chews with his mouth open is a real danger to a situation like this, or someone who blows their nose in their napkin at the dinner table. All sorts of interpersonal friction can build up.

"But we all know we're under a microscope from Mission Control and we try to be on our good behavior. In fact we repress our antagonism, our hostility, and this can lead to serious frustration.

"One way a lot of people find themselves coping also leads to serious problems, and that is to transfer the hostility to those outside our little world up here. We take it out on Mission Control, or our families, or even our scientific colleagues down below. The point is that you should be aware that this can happen. When you feel yourself surrendering to it, try to back away and put things into perspective.

"Now I started to cover the various stages or patterns that we see with long missions. About three-quarters of the way along you'll come out of your depression and suddenly become very elated; at least that's true of 90-day missions. Then, as you approach the final days of your tour you'll experience apprehension and the depression once again. Perhaps you haven't accomplished all that you hoped for, or you've been bitten by the euphoria of the space experience and don't want to leave. Whatever the reason, don't be surprised if you feel like this as you come toward the end of your voyage.

"I want to spend a couple of minutes on exercise and medical support. Both of these areas are very important, even critical to your well-being. One of the things that will happen to you here in orbit will be a certain amount of physiological deconditioning. We are a terrestrial species, we were conceived in gravity, and all of our lives have been spent in what we call a one-gravity environment. When we no longer have that pull on our bodies, it leads to certain changes in our body and in the way it operates.

"First thing your body decides to do to compensate is a shift in your body fluid from your legs up to your chest. This sends a message to your body to shed more urine than normal and that, in turn, results in a fluid loss, a drought, throughout your body's ecosystem. This deconditions your cardiovascular system, and with the reduced output from your heart, your heart does not beat as often, so your blood volume is cut down and your heart decreases in size.

"The rest of your body quite naturally has less of a load on it. This triggers a loss in calcium, nitrogen, and phosphorus, and a decrease in the size of your bones. Kidney stones form more easily, too. You'll also notice a decrease in your physical strength, and along with that there will be a loss of muscle tone. Your response time will suffer, and you'll lose some of your physical capacity for work. All of this is normal, so don't panic when you begin to notice symptoms.

"Now we have two bicycle ergometers and treadmills on board, and all of them are located next to windows, if you haven't already noticed. The old *Skylab* crews suggested we put them there so that you'd have Earth to look at while you exercise. There are also CRTs next to both of them so you can look at tapes if you'd like. And we encourage you to keep a daily log of what you do during each exercise period and what your heart rate gets up to.

"We believe in recreational exercise whenever possible. Back on *Skylab*, Pete Conrad rode his bicycle for 90 minutes once so that he could say that he had ridden it around the world. On the tapes we have the famous bicycle paths of the world, and that's always fun to play through a couple of times on a flight. You can also watch television at the same time, or a movie or whatever. But just remember, with this many people on board you need to ride when you're scheduled. We don't mind if you trade times with someone else, just don't let the bikes sit empty if you're scheduled to use one of them.

"Now, I want to warn you that with weightlessness there is no convection current in this module. This means that the body heat that you build up while you're exercising stays right there with you as you continue, and it can get pretty uncomfortable. So we've provided small fans at both locations, and we recommend that you use them during your workouts."

"You mentioned kidney stones a minute ago," Wayne said. "Just what can you do if we have any medical emergencies while we're up here? It's kind of frightening to think about being stuck up here if we have a heart attack or brain injury or whatever."

"You mentioned a couple of emergencies most people think about," Finnegan replied, "but there are others as well. Abscessed teeth, appendicitis, major burns, broken legs or arms, they all scare hell out of people when they stop to think about it. But so far, in over ten years of operation up here, we haven't lost anyone. And once we get into the deep-space voyages, we must have complete facilities and capabilities on board.

There's no turning back when you're halfway to Mars. We're even equipped to handle a mortuary procedure if it ever becomes necessary.

"Actually, we're in very good shape as far as our response to serious illnesses or injury is concerned. You may have heard about our on-board medical support systems already and what we have in the computer, but if you haven't, let me take a couple of minutes to bring you up to date. It should ease some of your concerns.

"Most of the time we don't have a licensed physician on board, and even when we do it's nearly impossible for any one person to recall everything there is to know about the human body and the current treatments. So what we need to do is to provide an on-board body of knowledge, as up-to-date as we can make it, of everything that is known on Earth at a given moment about disease, injury, and suitable treatments.

"Several years before *Friendship* was launched, a group of people within NASA and at the University of Florida worked to develop just such a system. Dr. Ralph Grams and Dr. Wick Hoffler were two of the primary movers. They called it the clinical practice library of medicine, the CPLM, an on-line biomedical computer library. In the DOCS file, as everybody on board calls it—that's data orbiting care system—we have a portable medical support system that permits us to access expert information instantly, the moment we need it. We update both the hardware and the software constantly."

"I'm not really sure that I understand how this thing works," Billy said, with a quizzical look.

"Well, it's really a decision support system. It is an awful lot like a piano in the sense that it represents a wide range of medical input and ideas that can be controlled by a number of keys. The information—actually all the world's known information about medicine at any moment—is in the system. With practice, just like with a piano, we are able to bring up the right alternatives that we can use to jog our minds and memories.

"We actually have thousands of years of medical experience and knowledge in a package no larger than a shoebox. We've indexed all of the major words in the various texts, and that permits us to have random access to any word in our database."

"How do you get the right information out of it and relate it to a patient?" Mary asked.

"One of the secrets is in what they call Boolean constructs of words. For example, Mary, let's say you come to me and say that you're not feeling well. You have pain and you're dizzy or you may even be vomiting. I obviously know that you are a female. Now the connectors between all of these symptoms are the words 'and,' 'or,' and 'not.' You're a female, not a male. Pain and dizzy or vomiting. These connectors permit us to do our linguistic algebra. They guide us in our detective work.

"This kind of exercise is what takes so very long to learn in medical school. It's one thing to memorize the body and its diseases or injuries, but it's quite another to link that with the human being standing in front of you in distress of one kind or another. So we conduct a search of all of the known medical knowledge here in the DOCS as it relates to your symptoms, using a search logic that we have learned by practicing with the system. As they say in medicine, 'To make the diagnosis, you must first think of the disease.'"

"Does it show pictures, too?" Wayne asked.

"It does better than that, Wayne. It not only shows stills in full color but it also now has the capability of showing us three-dimensional surgical techniques by means of holography. We can work in conjunction with ground medical centers and specialists with the DOCS equipment. So, although there may just be me up here right now trained in this discipline, at a moment's notice we can link up with the major medical facilities around the world and the top people in each of the fields.

"DOCS not only helps us with its infinite search capability to define medical problems, but it also then takes us through the critical treatment phase. And since sound is also very important in diagnosis, the system has a memory of all possible sounds that are part of any given condition.

"We've even held joint consultations with medical centers around the world simultaneously and have provided them with X-rays and other test results that we've performed up here. We really have the system down to a fine art and are fully qualified now for our Mars mission."

"When you pull up a portion of the text out of DOCS, I assume the pictures referring to that disease or injury are part of what you see?" Mary asked.

"Right, but we also have all the photographs in separate topic, anatomic location, and disease files. And when a microscopic slide or X-ray film is relevant, that's also noted as part of the parent picture."

"Let me see if I understand this now," Wayne said. "You first enter key words into the system and it in turn shows you the possible disease or injury alternatives that they could represent. Is that about it?"

"Basically," Finnegan replied, "although that is a bit simplified. But you've got the general idea."

"Wow, what an incredible system," Wayne said. "It's certainly reassuring to know we have DOCS on board. And it'll be interesting to see how we handle the various psychological stages that you mentioned when they come along."

"Just remember, living in three-dimensional space, in a 360-degree world, can take some getting used to. Take it slow and when things come up that disturb you, let's talk about them. Whatever you do, don't fret about them. Just tell your old MOM."

"Thank you, Finnegan. I'm sure we will," Wayne said. The others nodded in agreement.

"There are just a couple of other things that I need to mention while we're on the personal end of this thing up here. You're each going to be assigned a big brother, and, in Mary's case, a big sister. Edie will be your big sister, Mary. And, Billy, Barney Caldwell will be your big brother. Wayne, you're going to have to put up with having me as yours. We've found that this kind of relationship really helps new crew members over the rough spots and helps them adjust all the quicker.

"You'll find that several people on board on this tour are what we call androgynous. They're the type of crew members who have the capacity to be goal-oriented and still retain their sensitivity toward other people. They have good self-concepts and can very quickly build rewarding friendships. And since that's the name painted on the side of this critter, we've found the system very effective. Your big brothers and sisters all fall into this category, he said modestly.

"But, seriously, for some reason some of us handle this aspect of life better than others. So don't hesitate to come to us with your questions.

"As you settle in you'll notice how comfortable life on board is. A great number of people down at NASA/Ames and elsewhere did a whale of a lot to give you this feeling. Things like body motion envelopes, social logic, and a whole host of variables were considered in finite detail before the station was even begun. People in industry and universities as well as NASA agonized over the layouts of equipment and the like for years before construction began.

"Well, enough. I think I've covered most of what I had on my list to talk to you about, and if I think of anything else I'll know where to find you. You can't go far. The big thing is to develop an eagerness for the game and let your enthusiasm lead the way. I'm sure the three of you will do just fine."

13

Olympus Mons or Bust___

Stu turned to the assembled crew. "Let me get us on the network and then we can do our thing. Houston, *Friendship.*"

"Good morning, Stu. How are things upstairs today?"

"Pretty good, Dave. Have you got the net set up?"

"Just about. How's our picture this morning?"

"It's been better. The solar flares have really been playing havoc with our television the last couple of days. How you?"

"Yeah, we're having the same problem. I think Bert about has everything set up. Stand by."

It was changeover for the blue and red teams aboard *Friendship,* and they were all gathered in the wardroom for this historic occasion. For Mary, Billy, and Wayne, it was coincidental that they should be aboard at such a moment. But it would make the memories of their tour all the more special.

"*Friendship,* Houston."

"Go ahead, Dave. We're all here, ready or not."

"*Kalinin,* Houston."

"Good morning, Houston. And how are you?"

"Very good, Aleksei. Is it still snowing around Moscow?"

"Oh yes. It seems winter comes earlier each year. And is *Friendship* standing by?"

"We're up here, Aleksei. As a matter of fact, we have a surprise planned for you."

"Ah, that is good."

"Houston, Paris here."

"Hello, André, *tu as bien dormi?* How did you sleep?"

"Friend David, thank you. *Oui, j'ai dormi sur lex deux oreilles.* I slept soundly on both ears."

"Good, André, hold on."

"Tokyo, Houston."

"Ohio, Houston."

"And good morning to you, too, Watrau. Stand by."

"Montreal, Houston."

"Good morning, Dave. Fletcher here."

"Aleksei, do you have Peace Station on line, too?"

"Of course, Houston, have we ever let you down?"

"Good morning, *Mir,* welcome to the club."

"Hello from Peace Station."

"Well, now, we have everyone on line. Isn't technology grand. Before we get started, Aleksei, we all have a message that we would like to give you and all of our friends in the Soviet Union. Okay, everybody, don't screw up. Ah one, and ah two, and ah three . . ."

From around the world, from Tokyo and Paris, from Houston and Montreal, and especially from the orbiting *Friendship* came the slightly out-of-sync sounds of a familiar tune.

> Happy birthday to you,
> Happy birthday to you,
> Happy birthday dear *Sputnik,*
> Happy birthday to you!

Laughter and applause could be heard over the network when they were done.

"Ah, thank you everyone, you did not forget. It will make the Secretary General very happy when I tell him."

"Well, good, Aleksei. How could we forget today. It's not every morning that we get to celebrate the 50th anniversary of the little beeping voice you put up that started all this. I keep telling you, though, that Eisenhower gave you our timetable, and all you had to do was beat us to the punch."

"Ah, well, I suppose you are right. It all seems sort of silly now, doesn't it, especially now that all of us are so hard at work together on this Mars project. Well, on behalf of all of us here in Moscow, thank you again for remembering our special day."

"Well, that's not all to the surprise, Aleksei. Vladimir, do you read us okay?"

"Da, Houston."

"The rest of the surprise is that the schoolchildren in North and South America, the Orient, Africa, Australia, and Europe have all taken up a collection, and with their pennies they have bought a special gift for one another and for all of the young people in Russia. We hope that you can get us some of your special brand of clearance, Aleksei, because, on his way home tomorrow, Smokey is planning on paying Vladimir and the rest of our friends on board Peace Station a special visit to deliver it. Do you suppose you can arrange that with General Leonov?"

"Of course, David, of course."

"Good, because we thought that it would make the gift all the more special if we delivered it between our two orbiting space stations in honor of the special anniversary of little *Sputnik*."

"What is the surprise, David?"

"Vladimir, I think you've been up too long; you're getting impatient."

"Well, you know how much fun we always have when we exchange visits like this. We may even get out our forbidden vodka."

"Hold on, Vladimir, that's not fair. Everybody on board *Friendship* just might try to stowaway on *Discovery* and pay you a visit."

"The surprise, David, the surprise."

"My, we're impatient today, Aleksei. At any rate, our schoolchildren have had a golden replica of *Sputnik* built, to the exact size, I might add. And each country has had special scrolls signed by all of the children across each land and then microfilmed down. All of the names of our children are inside the *Sputnik*. After we deliver it to Vladimir, we want you to announce it to the children in your country from orbit. We can have a joint broadcast, if you'd like, of Smokey presenting it to Vladimir, and then when your shuttle pays its next visit to Peace Station you can take it back down to Mother Russia.

"We hope you will put it on tour in your country so that all of your children can see it and so that they, too, can sign the scroll of peace and friendship. Then we can put their names inside *Sputnik*, too. And we hope that it will then be sent on a perpetual journey around all of the schools in our separate countries so that it will always stand as a reminder of the good will that we all share. Do you think you can arrange that?"

"Of course, David. And thank you all. I think that is a fine idea. So much better, yes, than keeping it in a dry old museum."

"Aleksei, Stu here on *Friendship*. How are you today?"

"Very well, Commander. Especially after this touching gesture."

"Good. I just wanted to say that after we get over today's regular business you and I can have a short exchange with Vladimir and Smokey to plan *Discovery's* rendezvous with you all."

"Good."

"Paris, Houston."

"We're still here, Dave."

"Good, I thought we'd lost you there for a minute. Since you're this month's chairman of the committee, André, why don't you go ahead and get started before this solar flare business gives us all fits. Watrau, I'm sorry about your predecessor. And I know I speak for everyone when we welcome you to our small group. Did he make it through the surgery okay?"

"Yes, thank you, David Christiansen. It is an honor for me to be a part of this project."

"Please, everyone, this is André. Let's do what some of you call a round robin and each of us report on our progress since our last conference call. Aleksei, since this is your special day, why don't we start with you."

"Very well, everyone. Mother Russia gets to be number one this month."

"I think I hear an old Bolshevik somewhere on this network."

"Ah, well. A little competition never hurt anyone, as you Yanks like to say. So, where to begin. Our work on the Mars rover is on schedule. I think I reported to you last month that our new tires were working better. David, please thank your people for the lunar-module wheel technology. We made some improvements to the work your General Motors people did way back then, but it has been very useful to have all of your wire mesh designs. We revised the static load distribution in view of the increased gravity on Mars, but we kept the titanium bumper and springy rings inside the tires."

"Good, I'll pass the word along to Detroit and Santa Barbara."

"Next, the Venus landers. We have all of our Venera data from the seventies to build on, and that has been a big help to our Space Institute people. Our targeting people would like you all to consider the big volcanic crater on the eastern side of Maxwell Montes on the Ishtar Terra highlands. As some of you may know it is a huge caldera more than three times as large as anything we have here on our little blue planet. And, since Maxwell Montes is the largest mountain on Venus and is over a mile higher than our Mount Everest, we feel that it could shed a great deal of light on our understanding of both the dynamic lightning patterns we saw there before and of the geology of the planet. At least think about it, will you?"

"Of course, Aleksei," Dave answered from Houston.

"With all of the ozone studies we're doing from both *Friendship* and our own *Mir*, our Peace Station, we're very anxious to study the greenhouse atmosphere on Venus some more. A lot of our new instruments are going to be very useful on our probes. Our large probe has a nephelometer for cloud studies, a gas chromatograph, a mass spectometer and so on. I can get into more details later if you'd like."

"Aleksei, this is Fletcher here. How are you coming on the redesign of your habitation module to fit into the cluster?"

"Very well, Dr. Hawk. We have the new universal utility lines installed now up at Peace. Vladimir and his crew have been busy the past couple of weeks and we still plan on being on schedule for the mating of the modules in the spring. We've sent the latest changes to the life support system to each of you, at least they were supposed to have gone out yesterday. Let me know if you didn't get them yet. We are very close to 90 percent. I know you all said you'd be happy with 80 percent, but we're trying to reach our goal of 90 percent efficiency in the recycling. That will be a big plus when we factor it into our module cluster. I'm sure you'll all agree."

"That's excellent, Aleksei," Dave put in from Houston. "Stu's people at *Friendship* have just about completed—but why don't I let Stu tell you himself. Stu?"

"Thanks, Dave. Ah, we've completed nearly all of the assembly connections with Nodes 3 and 4 now, Aleksei. So that means that our hab module and the updated spacelab module are pretty well connected at our end of the cluster. The garden module is due shuttle after next, that is if your people got the new loop sent over to the Cape. Do you know if it left yet?"

"Aleksei, I can answer that. Yes, the Cape has it and they're installing it now in the integration and engineering building. Soon we hope to make up the time we lost on the new nutrient system. So we should be back on schedule with that lump pretty quickly."

"Thanks, Dave."

"André here, Stu. You said you have Nodes 3 and 4 almost done. Do you foresee any problem interfacing with the second stage propulsion system now that you're starting to assemble the ship?"

"It's really too early to tell on that, but we hope to have a better idea when the ESA/Japan module arrives. These things always look easier on paper than they do when we're upstairs here trying to fit all of the pieces together. How are your people coming, André, and I guess I should include Watrau in that question too?"

"Our joint module is just about ready. For everyone's benefit, we're

Assembly of the Mars manned mission spacecraft is an ongoing project in the vicinity of the space station. It is a joint project of the United States, the Soviet Union, the European Space Agency, and Japan.

giving it a subsystem verification next week. Everybody keep your fingers crossed for us that it all works in sync. Watrau and his team are due here in Paris on Monday. Are you still getting in on the eight-thirty Orient Express, Wat?"

"That's still our plan."

"Good. Well, who wants to report next?"

"Fletcher here, André. We've been having quite a flap in our press lately about this business on the Great Stone Face and the pyramids. Are any of the rest of you getting any heat about it?"

"None here in Moscow, Dr. Hawk."

"There have been a few editorials here in the States."

"Well, here in Canada there is a growing movement to insist that the photo targets include closeup stereo studies of Elysium so that they can be checked out. I'm afraid people see more to the face than those of us closer to the project do. And they could be right, you know. It could be man-made. We have our own strange faces on Easter Island, remember, and a lot of other unusual phenomena scattered around the world. Perhaps we should all try to keep an open mind about it. What do you all think?"

"I think it's ridiculous, but I've been wrong before," Dave said.

"Stu here, gang. If you want my two cents' worth, I'd give it real serious consideration. You could all save a great deal of problems in the future if you set the thing to rest once and for all. We all know that the strong Martian winds may account for the pyramids just as they have here on Earth near the Gilf Kebir plateau, and we have a Great Stone Face or two made by nature ourselves, like that tourist place in Vermont.

"I really think we should urge the target committee to include the site in our prime target list. Heck, I'd like to land there and hike up to the Great Stone Face and ask her a question or two. Could be interesting. But seriously, we really should look into it."

"Your request is noted in the minutes, Stu. Now, let's move on. Fletcher, will the new Mars arm be able to do all you thought it would do when you let the contract?"

"We have a great degree of confidence that it will. As you know, it is always difficult to simulate weightlessness in our ground tests of these arms, but the new elbow and wrist designs seem to be just what we need.

"While I'm on the net, I should also report to you that we have the storage module ready to be shipped to the Cape. The revised storm cellar barriers ended up being a bit heavier than we had all anticipated, but I'm sure you'll all agree that the added safety will be well worth the tradeoff. Better to be overly cautious than to lose a crew in deep space to the radiation in case Old Man Sun acts up again.

"Well, I think that pretty well covers it for Canada. As soon as we have our railroad clearances on the module delivery I'll let your people know, Dave."

"Good. Thanks, Fletch. Maybe I should fill you all in on the progress on the Mars excursion module, the MEM. Grumman and Rockwell are doing their usual great job on this one. The final design changes are all made and we did add the extra stay capability as we discussed several months ago. We also were able to fit the new science instruments into that part of the gumdrop. It's the same basic shape as our old Apollo command modules but quite a bit larger. Of course, it has to be since it must have both descent and ascent stages built into it.

"As you know, our delivery date is still a little way off since we have to fly the thing in Earth orbit for a shakedown cruise. But once we get the cluster all assembled up in *Friendship's* backyard we should have the MEM ready to mate to the front end."

"Aleksei, Dave. Would I be prying if I asked how your final data looked on your *StarEagle* test the other day?"

"Not at all; the rest of you might be interested in knowing too. I hope that this joint Mars mission is just the start of a series of unified efforts like this and that in the not too distant future we'll be able to ride *StarEagle* together out to Jupiter and Saturn. Don't suppose any of us will still be in the business, but maybe our children will be.

"We've very nearly got the fusion thing licked. We did get a lot of overheating at our maximum levels, but we sort of expected that and were already working on some design changes. When we do our second shakedown cruise next summer, we hope we'll get even closer to the final design. But to answer your question more fully, Aleksei, just as soon as we get our papers all written I'll send a set to all of you, minus the things that make our oversight committee nervous about getting out too soon."

"I think we all need to put extra effort on this funding thing in our parliaments," Fletcher said. "We don't want another Apollo-type fiasco where we develop all of this fantastic hardware and the wherewithal to explore the Martian surface and then just head back home and sit on our fannies for another generation. What a terrible waste of forward movement. We simply must establish a foothold on the red planet and continue to build upon that achievement."

"Gentlemen," André broke in, "perhaps we'd better move on to the rest of our agenda for this morning. We need to cover the computers yet that the United States is going to provide as well as the joint propulsion system. Let's move on."

Stu turned to the group listening in on the worldwide television link. "I think maybe we'd better all get back in harness. I'll monitor the rest of the meeting and if anything significant comes up I'll put it on the P.A.

Matter of fact, we'll have the tape of the meeting, too, if you'd like to listen to it later. Fred, you've been working on the Mars machine for quite a while, why don't you take our three short-timers over into the library and fill them in on what's what with the project."

"Will do, boss-man. You troops want to grab some coffee and join me there? I'll get my briefing book and meet you in a couple of minutes."

Mary, Wayne, and Billy were soon doing their favorite thing, looking out the observation port in the Ames Library.

"What do you think of all this talk about the Great Stone Face on Mars?" Mary asked.

"Well, like Stu said, the best way to shut everybody up once and for all is to check it out," Wayne said. "Frankly, I'm surprised that they're not pushing harder to have the MEM land there."

"Maybe once they get into Martian orbit and have a chance to really give the face the once-over with closeup photography they'll decide to do just that. There have only been the two pictures taken of it by the Viking orbiter," Billy added.

"It sure gives me the creeps," Mary said, hugging herself.

"Why does everyone look so serious?" Fred asked as he joined them in the small library cubicle.

"We were just talking about the Great Stone Face and the pyramids," Mary said with a small worried smile. "You don't suppose it really could be a manmade face, do you?"

"It wouldn't have to be manmade," Fred teased. "It looks an awful lot like a woman's face to me. But maybe we'd better forget that for a while and move on to the mission itself. That's what Stu wanted me to fill you in about.

"Actually, we've been working on various aspects of the Mars mission ever since *Friendship* opened for business. I think maybe Beth covered part of the life support system end of it with you a couple of days ago. We're getting a great deal of help now from Aleksei and his people at their Bio-6 chamber at Krasnoyarsk.

"One of our major goals on *Friendship* was to solve the bone decalcification problem. We were losing up to half a percent every month at the start, but diphosphonates seem to be a big part of the answer to that puzzle.

"We've also solved the storm cellar radiation protection system, thanks to the folks north of the border. And we think we've pretty well got the cosmic ray and solar radiation exposure problem taken care of.

"Our forecasting has improved a great deal and so has our sensor system. And we've reached our goals of developing the proper kinds of exercise for the long flight to maintain heart and muscle conditioning. Finnegan no doubt filled you in on the medical support system, didn't he?"

"Yes," Wayne replied, "he covered that quite thoroughly. It was fascinating to see the system work firsthand."

"Good. Those were some of the things that we had to solve between the moon era and the Mars epoch. A weekend trip up to the moon seemed like such a huge step before we learned to do it, just like our Mars flight does now. But before you know it, we'll be planning a trip out to one of those suspicious-looking planets we think we see out there."

"Exactly why are we going to Mars, Dr. Weber?" Wayne asked. "I know it seems like a pretty basic question now that the project is so far along. But you have to remember that some of us were born after it was all debated so heavily back in the eighties."

"Not a bad question, Wayne. Too many people forget to ask the basics and so never grasp the reality of a situation.

"As far as why we are going, there are a hundred ancillary objectives, but I think it was a fellow named Doug Blanchard down at the Johnson Space Center who put it pretty damn simply. He said that there were three basic questions we wanted to know about Mars. Number one, was Mars evolution cut short for some reason? You may know that it even had Great Lakes at one time. Why did it dry up and become the dust bowl that it now is?

"Number two, he said, was whether or not Mars could still develop into an Earthlike planet. There are a number of proposals on how we could make this happen. Of course, they'd take a couple of thousand years, but if it would eventually relieve the population problem on Earth, then a couple of thousand years in the long history of mankind would not be unreasonable to wait.

"And that is related to Blanchard's third big question. If Mars can be developed into an Earthlike place, will it support life at some later date? We can forget all of the details of exploration and science, Wayne, and just focus in on those three things. I think you'll see that it is, indeed, worth the cost and trouble of going on over to Mars and digging around to find out.

"We had a good window for getting over to Mars just a couple of years ago in 2003. Earth and Mars were very close and we were thinking about heading a manned flight on over. About every 2½ years Mars is in the right position for us to get a really economical flight. I think it was Jim Oberg who said that in terms of the velocity needed to fly to Mars it only takes a little bit more to get there than it did for us to get to the moon. The difference, though, is that the moon is only a couple of days away and Mars about 9 months.

"We were thinking about using two spaceships so that we would not have to decelerate and then accelerate, and we had quite serious talks with the Soviets about building one of them. We called it the Flights of the *Eagle* and the *Bear*. The American *Eagle* first would have flown out to the vicinity of Mars with an international crew and dropped off an American and a Russian to land on the surface of one of the Martian moons, either Phobos or Deimos. Then it would have returned to Earth. Phobos is only 17 by 12 miles big and it circles 3,720 miles from Mars. Deimos is smaller, about 10 miles by 6, and it's over 12,000 miles out.

"The plan was that they'd explore Phobos, set up the research equipment, and then rendezvous with *Bear* a month later. In other words, the Russian ship would leave a month behind *Eagle* and pick them up for the trip home on the free-return fly-by route. But we got all bent out of shape with one another on that Middle East thing again. At least we all agreed to fly our sample-return mission then. That was better than nothing."

Wayne interrupted. "How many different ways of getting there are there? You mentioned the free-return fly-by route."

"Well, eventually we'll have orbiting Mars ships that will circle the solar system like interurban trolleys. You'll ride one of the advanced shuttles up to the Earth spaceport and then board a transfer vehicle for the ride out to one of the libration points between Earth and the moon. You'll refuel there and head out for a rendezvous with the cycling spaceship headed to Mars. Your transfer vehicle will ride along in a hangar. The trip out will take five to seven months.

"You'll then take your transfer vehicle from the cycling ship to the Martian spaceport. There you'll board a Mars lander and go down to the surface. When it's time to head on back home to Earth, you'll reverse the procedure and catch another cycling spaceship headed back to the blue planet from the red one.

"These cycling spaceships will be like ocean liners and they'll have artificial gravity that can be varied during flight. When you leave Earth, the gravity will be one 'g' and then it will be decreased down to 0.34 'g' as you near Mars so you'll be acclimated to that Martian gravity. Then on your way back to Earth the reverse gravity change will be done. That will probably be the interim system that we'll use, unless we luck out and get *StarEagle*'s fusion system perfected sooner than we expect.

"The other routes include a conjunction-class mission. The problem with that type of flight is that when you arrive at Mars after your nine-month flight, the sun is in between Mars and Earth. And we just don't like the idea of being blind during the critical rendezvous and landing portions of a flight. I suppose we could solve this by positioning communication stations in such a way that we could talk around the sun, but that all seems like it could add more potential problems than it's worth. And that kind of a mission would have to stay at Mars for a year and a half.

"An opposition-class mission is also possible when Mars and Earth are close together, but in that case we'd only be able to stay 20 days or so."

"Can I interrupt you?" Wayne asked tentatively.

"Of course. What don't you understand?"

"This business of conjunction and opposition is confusing. If you ask me, somebody screwed up on the terminology. You say opposition is when Earth and Mars are on the same side of the sun together. Brother, that sounds like it should be conjunction. And when Mars is on the other side of the sun, then that would be opposition. Doesn't that make more sense?"

"Sorry, Wayne, but that's the way they named it. Here, let me draw you a picture of it. Conjunction is when Earth is on one side of the sun and Mars is on the other. We can't see the red planet then, but just before it goes behind the sun we see it setting in the west right after our sunset; we call it the 'evening star.' After it moves out from behind the sun it rises low in our east just before dawn; then we call it the 'morning star.'

"This all happens," Fred went on, "because both Mars and Earth orbit the sun at different speeds. While Earth takes 365¼ days to complete one full orbit of the sun, 595 million miles, Mars only goes about half of a revolution. Remember, it's farther out from the sun, by half again, and it therefore moves slower around the sun. One full orbit for Mars takes 687 Earth days, and covers over 890 million miles. Earth moves at 66,636 miles an hour around the sun, while Mars only moves at 54,108 miles an hour.

"Now about every two years, Earth catches up again with Mars and passes it as the two orbit the sun together. If you're a racing fan you might say that Earth 'laps' Mars. Every 25 to 26 months Mars is about 44 degrees ahead of Earth, and that's our best launch window. We sort of have to 'lead the duck' when we shoot for it."

"It sounds to me like we could've gone to Mars a long time ago, is that right?" Mary asked.

"As a matter of fact, Mary, we could have gone out there right after we conquered the moon. We had all of the technology and hardware developed then. All of the forms to build the equipment, all of the trained people, thousands of them. And then we just threw it all away."

"How could we do that?" Billy asked.

Fred just shook his head. "It didn't make sense, I'll agree. But that's what we did. At the moment of our finest achievement on the moon, we made our tackiest blunders. First it was the Vietnam War and then Watergate. But I guess that's all in the past. The important thing is that we now have a second chance. And by making it an international mission, we hope we'll finally be able to direct mankind's attention toward the stars and away from the mud."

"We've never been to Mars, physically, that I understand," Wayne said. "But I know we've been there intellectually for generations, what with our orbiters and landers. Is that how we know that Mars has a dark blue sky like ours at dawn and that it then turns to salmon pink at noon?"

"Yes, after the sun comes up the convection currents start to do their thing with the dust particles and the Martian sky turns reddish. Why don't I get into some of that science with you?

"Let's go back in time for a bit, shall we? Let's go back to when Mars was just a fuzzy rust-colored spot in the sky with strange patches on it. You know that some people thought they were manmade canals and that there were Martians that lived there. But as it turns out, there are no Martians. At least not yet. But you know, someday *we* will be the Martians.

"However, getting back to Mars History 101, we first started sending probes out there in the early sixties. Our *Mariner 4* spacecraft sent the first pictures back of another planet when it flew within 6,118 miles of Mars on July 14, 1965."

"I thought the Russians got to Mars first," Billy said.

"Well, technically, they did achieve the first soft landing with their *Mars 3* on December 2, 1971. *Mars 2* crash-landed just five or six days before. Actually, the *Mars 3* only transmitted for about 20 seconds before it went dead. It never did send back any decipherable pictures.

"Then our *Mariner 6* flew over the equator of Mars at an altitude of 2,120 miles on July 31, 1969, and sent back 75 television pictures at 45-second intervals. A few days later *Mariner 7* covered the southern hemisphere and the south pole area and sent back 126 pictures. Not too many people paid a lot of attention at the time, outside of the scientists, because we had some people getting an awful lot of attention from a trip they made out to the moon that same month in *Eagle* and *Columbia*. Then in 1972 *Mariner 9* sent us back 7,300 good closeup pictures.

"On July 20, 1976, exactly seven years to the day since the first astronauts landed on the moon, *Viking 1* dropped from orbit. It was late in the afternoon on Mars. There was a light wind from the east that switched around to the southeast after midnight. The temperature that afternoon on the Plains of Gold was 22 degrees below zero.

"Earth was 200 million miles away at the time of the landing. The first picture came into JPL's control center in Pasadena a couple of minutes later and showed the footpad of the lander and some angular Martian rocks. Viking's cameras were located about five feet off the ground, so the photographs that we have look like they were taken from the eye level of a person standing there."

"Where and when did the second Viking land? There were two, weren't there?" Wayne asked.

"Yes, there were. *Viking 2* landed 45 days later in an area known as Utopia Planitia, the Plains of Utopia. That's about 4,600 miles northeast of where Viking 1 was. Both of these landers were designed to operate for 90 days, but in reality they sent back good data for well over six years. A tremendous return on our science investment."

"Question?" Wayne held up his hand.

"Shoot."

"If Mars is only half the size of Earth, how can it have mountains that are twice as big or a canyon that is 13 times larger than the Grand Canyon? It seems to me I read somewhere that the Martian Mariner Valley would stretch all the way from California to New York and that it goes one-sixth of the way around Mars."

"Right. It's 2,800 miles long, 370 miles wide, and 4½ miles deep. It makes our Grand Canyon look like a feeder stream. And Olympus Mons is so huge it would cover our state of Missouri. Its base is 375 miles across and its crater alone is 50 miles across. Rhode Island could fit inside the cone. The mountain is 79,000 feet high. Mount Everest looks like a foothill next to it."

"But why is everything so much bigger on Mars?" Billy persisted.

"I guess that's one of the things we want to find out. Mars has a different gravity than Earth, for one thing; that probably plays a part. Then there is the very important fact that Mars does not have the crustal plates that Earth does. Planets have to be a certain size before they have the same combination of volcanism and plate tectonics that we have here on Earth.

"It's all tied in with the internal heat forces. When the planet burps and releases heat and molten rock, there aren't any plates to slide around like we have. So the same vents and outlets are used over and over. The mountains just keep on building up and getting bigger and bigger. Whereas here on Earth, our plates keep sliding around, and we tend to get chains of mountains forming out of the same sort of activity. The Hawaiian string is a good example."

"How about the Big Valley, Marineris?" Wayne interjected.

"You've got me on that one, Wayne. But if I had to guess, I'd say the fact that the Martian Great Lakes were once located there in the great rift valley around the equator had something to do with it. They could have been as much as three miles deep according to Steve Squyres and Susan Nedell, who did a lot of interesting research with some of the Viking photos.

"Since Mars is so cold it would have meant that these lakes were frozen over most of the time. Perhaps planetary thaws happened from time to time and the floods and winds that resulted helped further erode the rift valley. That's another of the interesting things we want to find out when we land there. We're also damn curious to see if we'll find any snail or fish bones in the walls of the valley. It should contain a complete geologic record of the planet's history just like our Olduvai Gorge and other places on Earth do."

"I didn't realize that Mars had that much water at one time," Wayne commented. "What happened to it?"

"We figure that there is still enough water on Mars to make an ocean 300 feet deep that would cover the United States and most of Canada. You have to remember that we're talking about a planet that is colder than Antarctica. So all the water is frozen. Some people even feel that we can drill down and tap this water source underneath the permafrost.

"Incidentally, the north pole of Mars is water ice; its south pole is mostly carbon dioxide ice. Matter of fact, we think that Mars may be in the middle of an ice age right now, just like we have here on Earth from time to time."

"You mentioned the winds earlier. How strong are they?" Wayne asked.

"Let's see, I think they pretty well stay under about 15 miles an hour under normal circumstances, but when Mars is closest to the sun in its orbit the dust storms swirl out of the southern hemisphere and can build up to 150 miles an hour or so. The tiny dust particles they pick up can cover the whole planet.

"An interesting thing that we discovered with our Vikings is that the Martian weather is rather bland most of the time. It has its seasonal changes, but it doesn't have the really sudden changes like we do here on Earth. The answer, of course, is that you don't have the large bodies of water on Mars and the large cloud formations like we do to cause convection currents and all the turmoil and boiling that can occur with heating and cooling.

"You know, there's another interesting thing that I should mention so that you can envision the Martian planet as a real place. And that's the fact that its southern hemisphere is really an ancient cratered surface like our moon. But the northern is geologically a lot younger, having gotten a newer surface from time to time from the lava flows from its volcanoes. Why this is so is something we need to find out."

"You started to tell us about some of the other spacecraft that went out to Mars in the late eighties and nineties," Mary said.

"Yes, I do tend to get off the track, don't I. Well, the first of those was the Soviet Phobos lander that arrived on their May Day in 1989. It was the first major attempt at obtaining new knowledge about Mars since our

Vikings. There were two landers just as we had had on Viking, but there were some new wrinkles this time.

"We were stuck in just the two landing sites with our Vikings. But the Russians kicked out two spring-loaded hoppers about the size of basketballs that sampled about ten different places out a mile or so beyond the location of the landers themselves. These things jumped around like Mexican jumping beans. They'd propel themselves to a new spot, wiggle, and right themselves like a chicken settling down on a nest of eggs, and then perform a bunch of observations at each location.

"They radioed their findings back to the lander, and, in turn, it sent them up to the orbiting mother ship for transmission back to us here on Earth. We got a wealth of new information out of them. The landers had television, of course, and seismic detectors, spectrometers, and all of the usual science machinery."

"Why didn't they land on Mars, instead of on its tiny moon?" Billy asked.

"Phobos gave them a good platform for studying a fair-sized portion of the Martian surface, and they couldn't have done that science if they were stuck down on its surface. They looked for water and moisture using a neutron moisture meter and studied the atmosphere looking at the ozone, oxygen, carbon dioxide, dust, and water vapor.

"We're all interested in knowing how much fuel we can manufacture out of the water, carbon, and nitrogen on Phobos. It could very well turn out to be a refueling stop for us for the trip back to Earth. You know, the gravity on Phobos is so low that we don't actually have to land on it, we just sort of have to dock with it.

"Shortly after that flight, we in the United States sent out our Mars observer in August 1991. It was in a Martian polar orbit some 224 miles up, sun synchronous, and it made an orbit every 117 minutes. The observer gave us a complete global study of the surface of Mars and its weather every 56 days. We got a good handle on what elements and minerals the surface is made up of, and we were able to map the gravitational and magnetic fields. We also did a great deal of climatology science. Finding water was a big goal.

"Then the big one was when we flew our joint unmanned sample return mission with the Soviets, the Europeans, and the Japanese. We called that Operation Thoreau."

"You mean after Henry David Thoreau? I don't understand. Wasn't he a philosopher?" Mary asked.

"Oh, yes. But he was also one of the first to get us all to pause and realize that there is much more to life than just the acquiring of material

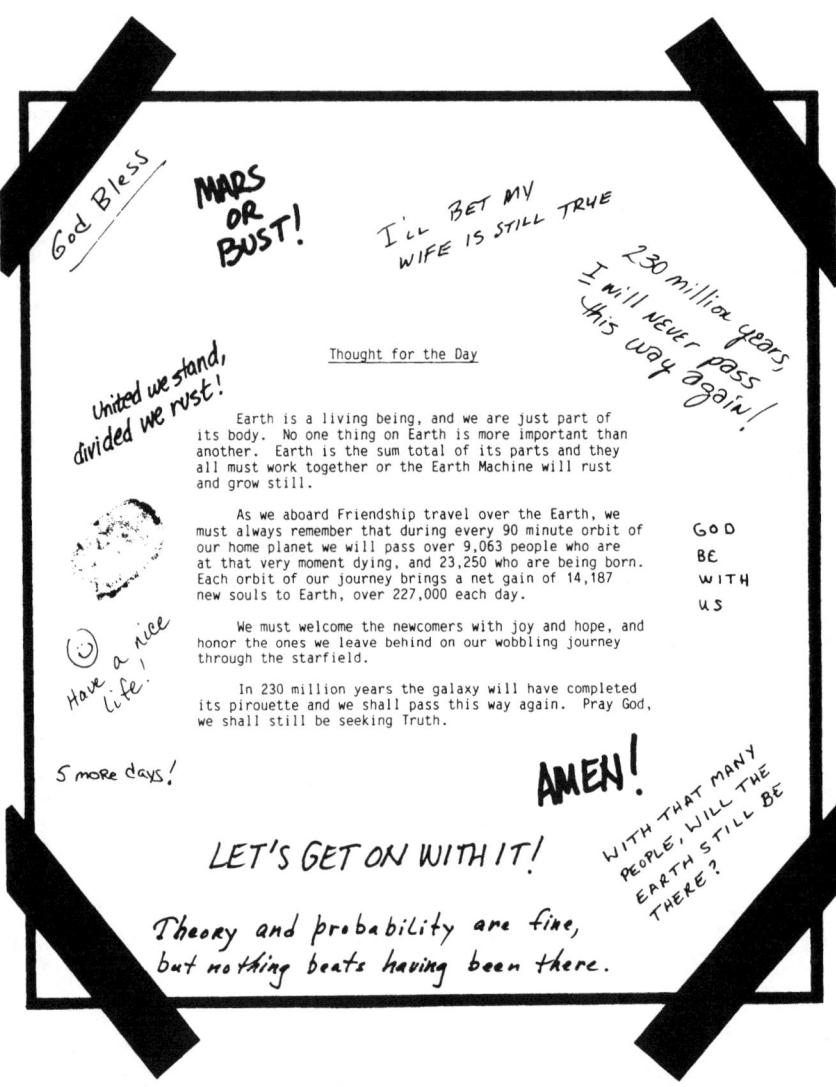

God Bless

MARS OR BUST!

I'LL BET MY WIFE IS STILL TRUE

230 million years, I will never pass this way again!

United we stand, divided we rust!

Thought for the Day

Earth is a living being, and we are just part of its body. No one thing on Earth is more important than another. Earth is the sum total of its parts and they all must work together or the Earth Machine will rust and grow still.

As we aboard Friendship travel over the Earth, we must always remember that during every 90 minute orbit of our home planet we will pass over 9,063 people who are at that very moment dying, and 23,250 who are being born. Each orbit of our journey brings a net gain of 14,187 new souls to Earth, over 227,000 each day.

We must welcome the newcomers with joy and hope, and honor the ones we leave behind on our wobbling journey through the starfield.

In 230 million years the galaxy will have completed its pirouette and we shall pass this way again. Pray God, we shall still be seeking Truth.

GOD BE WITH US

Have a nice life!

5 more days!

AMEN!

LET'S GET ON WITH IT!

WITH THAT MANY PEOPLE, WILL THE EARTH STILL BE THERE?

Theory and probability are fine, but nothing beats having been there.

One of the first things that all newcomers aboard space station *Friendship* see is this message that an earlier crew member taped on the wardroom wall. Over the years other crews have added their own comments. In so doing they have made it their personal statement of purpose and hope aboard *Friendship*.

things. Walden Pond helped bring Thoreau close to the universal aware-ness that we all are trying so hard to attain today. We hope that one day we'll have a Walden Pond on the surface of Mars, that we'll be able to bring life to that dead planet."

"I'm beginning to get the idea," Mary said, "that a lot of scientists think that Mars can be a new Earth."

"Yes, there is a lot of speculation along those lines. You know, we now have over 6 billion people on Earth, and by the year 2050 we'll have 10 billion. We're simply going to have to look elsewhere for our natural resources and for new territory to expand into.

"Mars can give us a permanent entry to the asteroid belt and to the outer planets. I think it was NASA's James Beggs who said that the universe doesn't care who explores it. He said that it would give us all on Earth something other than armed conflict to focus on and that, ironically, Mars, the symbol of war, could in reality become a powerful instrument for peace."

14

The Shining Princess of the Young Bamboo_____

"They're off in their own areas working right now," Stu said, "but later today when they wander back over here for dinner I'd like you to spend some time with the crews of the ESA and the Japan modules."

"You know, I really hadn't thought about it before, but do you all eat and sleep in the same modules?" Wayne asked.

"Oh, yes. We decided a long time ago that any other approach to this international station would be foolish."

"Do you ever have any problems?" Billy asked.

"Of course we have problems," Stu laughed. "What big family doesn't have problems? But we try to keep the conflicts to a minimum, and we try to understand and to respect one another's backgrounds, habits, and cultures. Once in a while we have to remind ourselves that in the year 3007 people will probably consider this a prehistoric spaceship. That in itself is pretty humbling, and it helps keep the egos in check."

"You said there were a bunch of countries involved when this whole thing was put up here," Wayne said. "Who were they?"

"Well, in addition to the United States, there was Canada, Japan, France, West Germany—actually they call it the Federal Republic of Germany—the United Kingdom, Italy, Switzerland, Spain, Sweden, Denmark, Belgium, Ireland, Norway, Austria, and the Netherlands. Finland also came into the fold as we went along."

"That's quite a big chunk of the world, isn't it?" Mary said.

"Well, it was a big job. But it wasn't all peaches and cream when we first started planning *Friendship*. Matter of fact, ESA almost pulled out. One problem after another came up, and before you knew it there was friction. They argued about who would operate the station, who would be located where, who would pay for what, and on and on. Sort of like a blindfolded committee designing an elephant. But eventually both sides decided to compromise a little."

"Didn't I read somewhere," Wayne put in, "that the degree of micro-gravity played a part in it?"

"Right. The location of the U.S. modules on the station gives them the best microgravity. They settled the argument by setting up a procedure so that everyone could do certain of their most gravity-critical experiments here in the U.S. end of *Friendship* and the rest in their own modules. For one thing, the work they do in their own modules is easier to keep secret so that the intellectual integrity of each nation is protected. And that's worked out very well."

"Were there any problems with sizes of the different parts of the station, what with the metric system and all?" Mary asked.

"Oh, yes. The ESA module, for example, is almost ten inches smaller in diameter than the U.S. modules. But the racks that fit into all of the modules are the same size and are interchangeable. They can be moved from one module to another through the common hatches and all. But listen, why don't I give you a quick briefing on each of the international portions of the station? Then when I take you around to see them later you'll have a leg up on what you'll be looking at."

"Sounds good to me."

Stu began with the European module, Columbus. The name was ap-propriate since Columbus came from Italy by way of Spain and since it was the 500th anniversary of Columbus's discovery of America when they started to assemble *Friendship*.

ESA used their Spacelab for the basic concept for their Columbus module. Spacelab, however, was designed to go up occasionally for per-haps a week at a time. To make Columbus permanent, the power had to be increased, as well as the cooling capability. And the overall length from the old spacelab configuration was increased for more volume.

Aeritalia, the ESA Columbus module contractor, put together four Spacelab segments instead of just two. The same basic module with a resource section attached was also used as a man-tended free flyer out a

The Columbus module, the European Space Agency (ESA) portion of the space station, was named in honor of the 500th anniversary of Columbus's voyages to the New World. As we all know, it was originally scheduled to be launched in 1992, but that date slipped several years because of the delays in the start up of the station after the *Challenger* tragedy.

ways from the space station. It was serviced from on board *Friendship* as well as reloaded and updated from time to time. Then it was cut loose, and it began to fly in formation with the space station in an unmanned mode.

The free flyer's gravity environment was one order of magnitude lower than it would have been if attached to *Friendship*. The integrity of the gravity environment on *Friendship* was affected by movements on board, as well as by docking and undocking the orbiter, the other free flyers, and the OMVs. Some of the critical experiments and production processes could not stand that sort of gravity pollution. That's why the free flyer was essential.

There was both interior and exterior access to the module shell for inspections and repairs when necessary. Columbus was designed to be compatible with the *Ariane 5,* its Hermes space shuttle, and the European data relay satellite.

Columbus used hundreds of times more computing power than the old spacelab. As *Friendship's* computer systems were designed, it was necessary to anticipate the different kinds of software and hardware that would come along in the twenty-first century. Commercial software was used whenever possible rather than designing space-specific software, and engineering design systems had to be standardized across Europe.

"Is the Columbus module all there is to the European part of the Station?" Billy asked.

"Oh, no," Stu said. "Actually their Eureca free flyer was put up along about 1991, about a year before work was started on *Friendship*. It was a platform in a stable orbit serviced by the orbiter. Then you all may remember when the Columbus unit itself was brought up here in late 1994. But, before I get into that, you should know that we now have a new enhanced Eureca platform in a co-orbit. It is also used for microgravity experiments and space-science payloads.

"The polar platform came up in 1995—I forget the exact month—and then in 1996 the man-tended free flyer. So those four elements actually make up the European side of things. Remember that *Friendship* is sort of like a base camp up here, but we have little spike camps spread out all over the place that tie into it."

"You said that they do some of their own microgravity things in Columbus as well as in the U.S. modules because of the difference in gravity as you move out from the centerline of the station. What else do they do?" Mary asked.

"Oh, the same sort of material sciences that we do in our modules, along with fluid physics and life sciences experiments. But in many cases they'll be working on things that are considered confidential. We've all

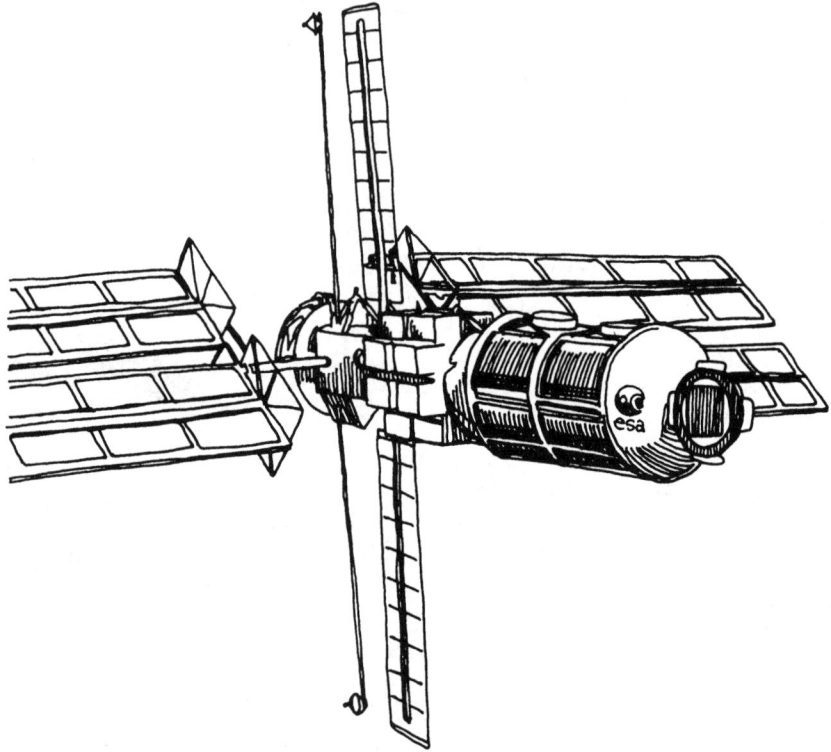

An ESA free-flying platform for experiments that need a pure gravity-free environment away from the tremors aboard the space station.

learned not to ask too many questions at dinner about exactly what each of us is working on in our modules.

"You may know that we even have a secured area in our materials module so that we can keep certain products and processes secret from one another. I suppose someday we'll have to contend with industrial spies on board, but I hope that's a long way off. Of course, once the European station is finished about five years from now, that problem will sort of take care of itself.

"Did I mention that the man-tended free flyer was launched up here by the new *Ariane 5*? It's a heavy lift vehicle just about like the ones made in the United States that launched most of the modules for *Friendship* in the first place. It can also bring up more modules and satellites. And you all know that it is also the launcher for the Hermes, the French space shuttle."

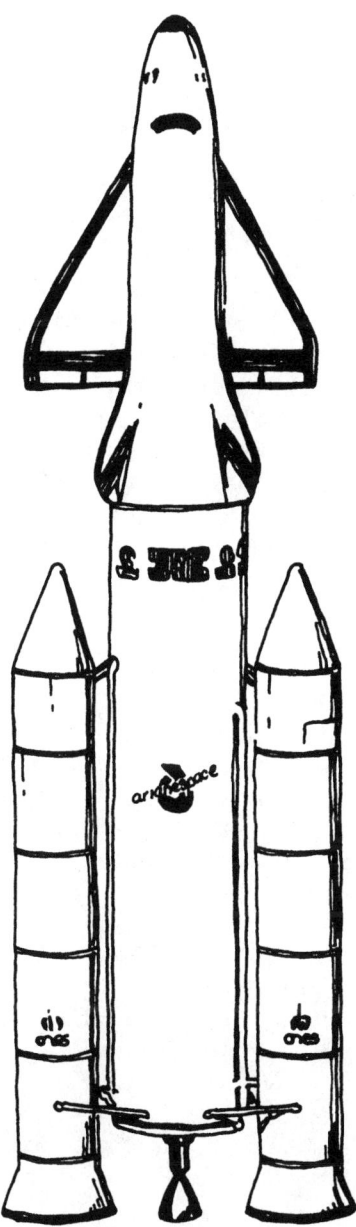

The French space shuttle, *Hermes*, is launched aboard the improved *Ariane 5* rocket system. It carries a crew of six astronauts and can stay up for a month on a free-flying mission. This ESA-built shuttle is only 59 feet long; the U.S. shuttle is 122 feet. It is a very sporty pickup truck design, as compared to the U.S.'s "dump truck."

"I've never seen a Hermes up close," Billy said. "Are we likely to while we're here?"

"Ah, let's see, I think so. I forget exactly what the launch schedule is, but I think one is due in three or four weeks. It's quite a bit smaller than the U.S. space shuttle, really more of a sporty pickup truck compared to our dump truck design. Our payload bay is 60 by 15 feet and the French one is 16 by 10 feet. I keep saying the French one; actually, it was built by ESA but it was funded by the CNES."

"What's that?"

"CNES, the Centre National d'Etudes Spatiales — that's the French space agency. Hermes was built by Aerospatiale and Dassault-Breguet about ten years ago. It's a hypersonic glider design and carries a maximum crew of six astronauts. It also lands on a conventional runway like the U.S. space shuttle does. But, unlike our original shuttles that could only stay up a week or so, Hermes can stay up for a month on a free-flying mission, and for three months docked here to *Friendship*. Of course, we have our own long-duration shuttle changes incorporated now, too, so the two craft are pretty compatible.

"Originally there was a lot of debate within the European Space Agency as to whether to build the Hermes or to try to make do with the British Aerospace/Rolls-Royce HOTOL. But I'll tell you, when they fine tuned the HOTOL so that it would reduce launch costs up to 80 percent, that was all the ESA people needed to hear and they voted to go for both systems.

"We Americans sure learned the hard way that you shouldn't put all your eggs in one basket. When our *Challenger* blew up, we had damn near done away with our backup expendable rocket systems. We sure had to scramble to catch up. So I think it was a far-sighted decision by ESA to opt for both the HOTOL as well as Hermes."

"We keep hearing it called the HOTOL on the news broadcasts all the time, but I can't remember what HOTOL stands for," Wayne admitted.

"Horizontal takeoff and landing. The HOTOL takes off from a regular runway — well, not exactly. It's launched from a laser-guided trolley-type system rolling down a runway. You probably know that HOTOL is usually flown unmanned.

"Did I mention that the HOTOL is all reuseable? It doesn't throw away its parts like the U.S. shuttle and Hermes do. It's a single stage to orbit vehicle."

"I'm not sure I understand how that's possible," Mary said.

"The secret is in the dual-functioning engine," Stu answered. "The thing breathes regular air from the atmosphere during takeoff and then switches over to its own internal fuel supplies. It only costs about one-fifth what our shuttle costs to get up here to *Friendship*. And, since it's just about the same size as the old Concorde, it can carry quite a load.

"I started to tell you about the secret to the thing. Back when it was under development in the late eighties there were a hell of a lot of people in Washington who were anxious to find out exactly how its engines worked. I mentioned that it carried its own liquid hydrogen, or did I? At any rate, it uses it with the atmospheric oxygen when it first takes off. It climbs at 24 degrees and after 2 minutes or so is supersonic. Within some 4½ minutes it has climbed through the commercial air lanes above 40,000 feet. And after only 9 minutes, it hits a speed of Mach 5 at around 85,000 feet.

"At that point it can't breathe regular air any more and it then switches over to its on-board liquid oxygen and liquid hydrogen for rocket propulsion. Its usual mission duration is about 50 hours before it has to return to Earth.

"We expect to see a HOTOL fly from Australia to Europe with about 60 passengers before 2010, just three years from now. Flight time will only be about 60 minutes. Actually, 67 minutes if you include time for air traffic control. Real flying time will only be 45 minutes or so. If you can afford it, you'll be able to eat dinner in Paris and have your after-dinner drink in Sydney."

"Holy smoke."

"Well, I'm not so sure it'll be holy, but it sure will be smoking along."

"Do we ever have all three up here at one time, the U.S. shuttle, the Hermes, and HOTOL?" Wayne asked.

"Matter of fact, no. That would be pushing our luck. We don't need that kind of traffic control problem up here in our little control tower. We do overlap two vehicles from time to time, but it's almost always the two shuttles, the French and American.

"You have to keep in mind that every docking and undocking is done very, very carefully and we have a great number of free flyers in the vicinity. To say nothing of the crew members that go out on a regular basis for one reason or another. Generally, if we even have a vehicle here, it's just one of the three."

"What about the Japanese module? You haven't said anything about it yet."

"Ah, yes. The module of the Shining Princess of the Young Bamboo."

"I beg your pardon," Wayne quipped.

"That's an old Japanese fairy tale that goes back to 900 A.D. or so," Stu said. "When we get over to their module later, remind me to point out the painting that they have in their library. It's 'The Tale of the Bamboo Cutter,' by Itcho Hanabusa. It shows the Shining Princess."

"Aren't you going to tell us the fairy tale?" Mary pleaded.

"Well, I really shouldn't spoil their fun. If I do, will you promise to act surprised if they tell you when we get there?"

"Sure we will."

"Well, this is a story about how the Japanese looked at the universe in ancient times. It seems that a long, long time ago in Japan, an old bamboo cutter and his wife lived alone. They desperately wanted to have a child, but just couldn't have one. And the bamboo cutter and his wife were no longer young.

"One evening after a long day's work, the old man saw a strange glowing stalk of bamboo in the twilight. He very carefully opened the bamboo and found a tiny girl inside, no more than three inches tall. Having no children of their own, he and his wife decided to bring her up as their own child.

"In only three months she grew into a beautiful young woman. And because of her radiant charm, she was named the Shining Princess of the Young Bamboo. Word of her beauty spread far and wide, and many suitors tried to win her hand in marriage. But she turned down all the eligible suitors, one of whom was the Emperor himself.

"Finally, on a night of the full moon, she ascended into the heavens escorted by her handmaidens, telling her parents that, in truth, she was from the Palace of the Moon and had to return. And that is the tale of the Shining Princess of the Young Bamboo."

"Why, that's a beautiful fairy tale," Mary smiled.

"Yes, it really is, and wait until you see the painting on silk cloth. Of course, the cloth had to be specially treated before we'd allow it up here, but that doesn't diminish the beauty of the Shining Princess. And if you think the bonsai trees that we're working on in our garden are something, wait until you see the ancient one that they keep in the JEM."

"Can you tell us more about the JEM?" Billy asked.

"I'll be happy to; I was hoping you'd ask. Basically, the JEM, the Japanese experiment module, is a two-person facility. It was brought up here along about the sixteenth launch during assembly of *Friendship*. It's actually a combination of several different scientific units.

"The pressurized module is 35 feet long and is used primarily for materials processing and life sciences tests. Attached to it is an experiment logistic module with the expected acronym of ELM. It operates pretty much on a 6- or 12-month cycle, depending on what they're working on at any particular time. It holds experimental samples, various materials for processing, and the usual gases that they use on board.

"It's returned to Earth when necessary, and then another unit is launched back up here from the Tanegashima Space Center, about 1,000 kilometers southwest of Tokyo. That's their largest launch complex. And the ELMs are sent up from the Osaki launch site there. They have two downrange stations that we sometimes interface with—Ogasawara and Christmas Island."

"Do they use the new H-2 launcher for the ELM?" Billy asked.

"Right. It can carry a payload 12 by 32 feet into orbit. It really gave Japan a workhorse launch system when it was first available back in 1992. As a two-stage rocket it can easily carry space probes to Venus and Mars, thanks to its restart capability on the second stage. It's opened up a whole new world of planetary science for them. The Tsukuba Space Center outside Tokyo has a pile of projects underway that they're working on now, thanks to the H-2.

"Another part of the JEM is the EF, the exposed facility. I've always thought of the JEM as sort of a grasshopper design and the exposed facility as the front end of the critter. It's where all the good stuff is located for the space environment studies that the Japanese are working on.

"They have equipment there for their cosmic ray burst observations and high-energy cosmic ray experiments as well as their large antenna system and a radio frequency interference unit. There's an airlock at one end that they use to move things like experiments, test pieces, and ORUs back and forth. The airlock is 50 inches in diameter and recovers 90 percent of its air when it's used.

"The other major part of the JEM is its manipulator arm. It looks like a grasshopper's leg. It's used for attaching and detaching orbital replacement units as well as the ELM. It has two different configurations for its arm. One is for lifting bulky payloads and has an end effector on it. The other has a small, fine arm for more delicate outside work."

"So they live here in *Friendship* too, and work over in the JEM?" Wayne said.

"Right you are. If we have any kind of a problem, though, they have a safe haven capability, usually up in their logistics module, the ELM, where they can last for up to two days while they wait for us to rescue them. Of course, their primary safe haven is here with the rest of us in the main habitation modules.

"You'll like Mitsue; she really has an infectious laugh, and she's one fine scientist. Shoji is pretty quiet, but he can be a very good friend once you get to know him. This is his third tour of duty. Between flights he's head of the materials sciences office at their astronaut training center there at Tsukuba."

"Well, you've saved the best for last, I see," Billy put in.

"Now, Billy, did you think I'd forget your home? If it wasn't for Canada we'd have had one heck of a time putting all of these Tinker Toys together, to say nothing of all the help your Canadarm has been on our space shuttles for the past 25 years.

"What Billy wants me to be sure to cover with you is our mobile servicing system. It's a descendant of the Canadarm remote manipulator system that we still use on the shuttle. The people at Spar Aerospace and the National Research Council of Canada built the thing, and NASA provided the transporter on which it moves along *Friendship*'s trusses. It was one of the first parts of the station to be brought up, since we needed it to assemble the entire thing.

"I know you've already seen it in use, and you'll see a great deal more of it as your time on board goes along. I don't believe we could get along without it. We use it for our regular maintenance program for the station as well as for moving equipment and supplies around the exterior. It also gets a workout deploying and retrieving satellites and other payloads. Sometimes we even use it in docking the shuttles.

"But one of its most valuable uses is in supporting the EVA activities of our outside crews. I think you can see it through the port over there. The large arm is 49 feet long; it's *Friendship*'s remote manipulator system. The shorter of the two is the one we use for an astronaut positioning arm; it's almost 20 feet long. Both have a seven-degree-of-freedom design, and both of the arms can be controlled either from in here or out at the EVA work station itself."

"I've heard a lot about the positioning system that the NRC developed up in Canada. Exactly what is that?" Mary asked.

"It's a super technique. It's based on a real-time single camera photogrammetry, and it gives the exact location and orientation of a target object 30 times every second. Remember that when we're working outside, or even when we're using the mobile servicing system from in here, we're faced with a wide variety of lighting conditions. Don't forget we have either a sunrise or a sunset about every 45 minutes.

"We also have a serious lack of reference points when we're trying to move something around out there. That can really give us heartburn at times. It's difficult to accurately gauge velocities and distances, and that's crucial when we're handling large payloads—especially during berthing operations—and when we're changing logistics modules, like we did the other day.

"The space vision system reads a group of target dots that we put on each object ahead of time. Actually there are four of them. The computer measures the position of each dot with respect to a known reference point. And, by using geometric relationships, the computer can also tell if the object is moving up or down, sideways, or even rolling and to what degree. It sure has saved us a lot of grief over the years.

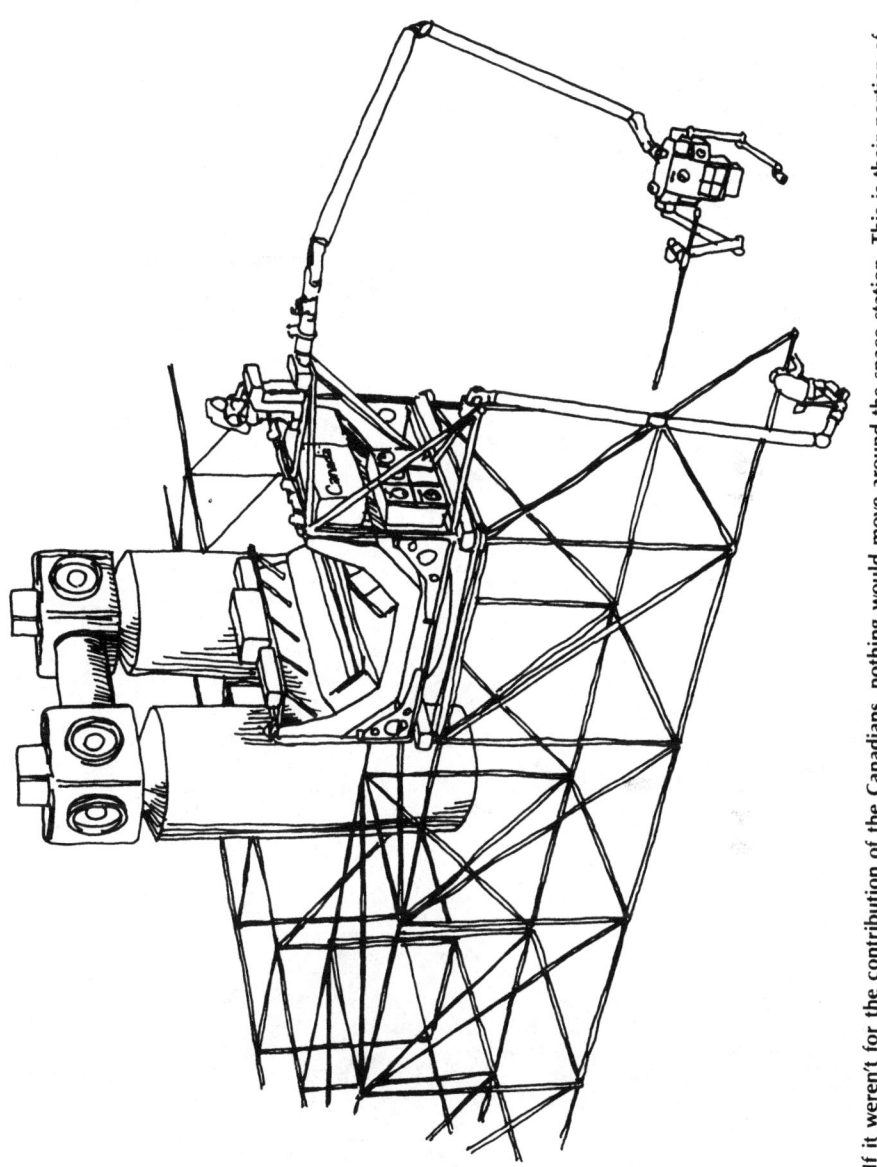

If it weren't for the contribution of the Canadians, nothing would move around the space station. This is their portion of *Friendship*. It consists of a mobile servicing center that moves all over the structure of the space station. Employing a unique space vision system to precisely locate items, it includes two remote manipulator arms. One, 49 feet long, is used for the movement of large units; the 20-foot arm is used as an astronaut positioning station for outside work activities.

"Well, that's a thumbnail sketch of the international side of *Friendship*. I'm sure that once you get over and look around the JEM and the ESA modules, you'll spot other things that I've forgotten to cover. Don't be afraid to ask questions. If you ask anything sensitive, don't worry about it—you just won't get an answer.

"That just about takes care of everything that I was supposed to cover with you on your orientation. Any questions?"

"Oh, yes. You said it yourself somewhere along the line. If we didn't all have questions there wouldn't be any reason for *Friendship* to be up here. To tell you the truth, I'm anxious to see that painting of the Shining Princess of the Young Bamboo. You've got my curiosity up," Mary said.

"Okay," Stu laughed. "Let's go meet her and then you people had better get down to work."

Epilogue

July 2019

Holding his mother's hand, six-year-old Michael Sean O'Sullivan stood at the foot of the *Apollo 11* lunar module on Tranquility Base. Standing straight and tall nearby was the same American flag that had fallen over into the lunar dust 50 years before as Neil Armstrong and Buzz Aldrin blasted off the moon's surface in the ascent stage of *Eagle*.

Sean's parents were members of the moon-based Columbia Scientific Community. Because Sean was not only the first child born at the new American moon base but also the first child born to Earth parents off the blue planet, his birth had sent a surge of pride through NASA.

Now, Sean and his parents had been invited to the ceremony celebrating man's first landing on the moon. It was a special day for everyone on the moon and on Earth.

Dr. Mary Two Hawks O'Sullivan, Sean's mother, was to say a few words to her colleagues gathered on the moon under the huge, clear protective dome. As she walked nervously toward the microphone, Mary touched the little buckskin bag that nestled under her tunic. A blue-and-white Earth rose slowly over her shoulder.

*   *   *   *

Far away in space the crew of the Mars-bound freighter *Alan B. Shepard* was watching the satellite transmission of the World Light Heavyweight Championship from Caesar's Palace in Las Vegas. The Astrophysics Officer, Dr. William Wong, slapped the Medical Officer on the back with a gleeful shout.

"Hey, Bob, he's coming back. The body work is starting to pay off. Bet you a buck he gets him down in the next round! Hope we don't lose the picture link again from Houston. This is the most excitement we've had around this old tub since we left."

The *Shepard* was one of a fleet of starships that routinely ferried personnel and supplies out to the red planet. Now that they were well into their voyage, the excitement of Billy's first tour out had begun to wear off. After the fight he was looking forward to the transmission from the moon, and to hearing his old friend, Mary Two Hawks, again.

As Billy waited for the transmission from the moon, he remembered his brief stint aboard the space station *Friendship* twelve years earlier with Mary Two Hawks and Wayne Morrison. Their few weeks together had created a bond that would last forever, no matter how far away from each other in space they might be.

The three had gone their separate ways after *Friendship*. Life on the space station was just one of many challenges they had met successfully. And Billy thought *Friendship* had a lot to do with it. Their visit gave them a vision for their own lives, their planet, and their future in space.

Billy smiled as Mary's face appeared on the screen. She hasn't changed, he thought. Life on the moon must be good for her.

*   *   *   *

In Tokyo, Vienna, and London, in Mombasa, Freeport, and Chartres, and in thousands of little one-street villages across the world, three-dimensional televisions were tuned to the ceremony on the moon.

In Brisbane, Australia, a middle-aged cancer researcher looked up from the paper he was writing for the San Francisco Medical Convention next month. Dr. Wayne Morrison had taken a position with the Australian Cancer Institute several years after completing his pioneering electrophoresis work aboard the old space station *Friendship*.

The venerable station still circling the Earth had nearly been swallowed up by all of the additions to it over the years. But, when its orbit was just right, its starlike brilliance still shown high in the nighttime Australian sky.

Wayne smiled to himself, filled with warm memories of his only spaceflight. Although the voyage hadn't been entirely smooth, overall, it had been rewarding. The valuable research each of them had accomplished aboard *Friendship* had hurled them to the forefront in their respective fields. Wayne wondered, though, how he would have made it without Mary and Billy and the support of the veteran crew.

Watching Mary Two Hawks O'Sullivan step to the podium, Wayne saw the dusty lunar module *Eagle* resting behind her. But it was the small boy by her side that caught Wayne's imagination. The boy was so at home on the moon. Wayne marveled. He wondered where this child might some-day journey and what distant thresholds he dreamed of crossing.

Photo Album

Friendship's Early Days

As we look forward to our joint mission to Mars and the continued successful testing of our *StarEagle* fusion-powered spacecraft, we should not forget the important pioneering work back in the 1970s, 1980s, and 1990s that got us this far. Here, then, are some of the historical photographs of those early days when space station *Friendship* was new, as well as from its predecessor flights aboard our first space station, *Skylab*. Our Apollo-Soyuz mission with the Soviets in the 1970s also set a precedent for our coming international voyage out to the Martian surface. In order to define the future, we must first understand our past.

Our first U.S. space station, *Skylab*, completed 35,000 orbits of Earth and was home to three crews during stays of 28, 59, and 84 days. This end-on view, showing the docking port, was taken during the final fly-around after the 84-day mission. On the right side, note the missing solar panel that was lost during launch on May 14, 1973. This was the last close look at *Skylab*. It fell to Earth and burned over Australia on July 11, 1979. *Photo courtesy NASA.*

American and Soviet spacemen exchanged tree seeds from their respective countries during our first international flight, the Apollo-Soyuz mission of July 15–24, 1975. The American seeds, white spruce from Wisconsin, were later planted in Cosmonaut Park in Moscow. Experimental tree seeds from around the world are also on the manifest for our joint flight to Mars from *Friendship's* vicinity. On the homeward-bound leg, they will be planted in soil from the Martian surface. *Photo courtesy NASA.*

The solar dynamic power system 50-foot parabolic reflector focuses sunlight to heat a fluid, which turns a generator to provide electrical power for the space station. Solar array panels are also used. Initially, this hybrid system provided 75 kilowatts of power, 25 kilowatts from the photovoltaic system and 50 kilowatts from the solar dynamics system. Nickel-hydrogen batteries are used to store energy for the photovoltaic system. *Photo courtesy NASA.*

This is what *Friendship* looked like when it was first built. A space shuttle approaches from below the station, while two crew members inside the cupola, atop the right-hand resource node, control the Canadian-provided remote manipulator arm. *Photo courtesy NASA.*

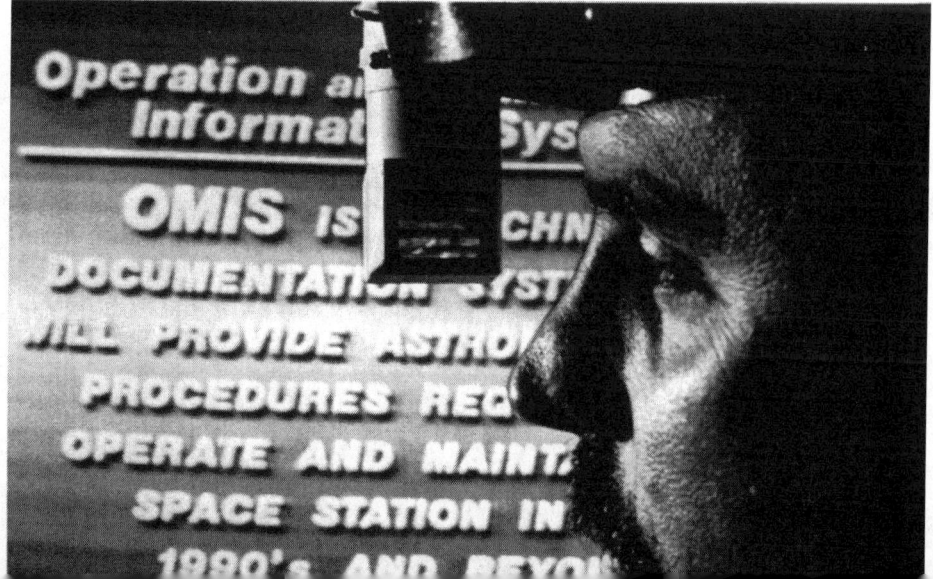

In this photo, which shows how we added to *Friendship* after it was built in the mid-1990s, the astronaut is moving a new connective node into place using the robotic Canadarm. By allowing another pressurized module to be added to the station, the node expands living and laboratory space. It was brought up to *Friendship* on the space shuttle, which can be seen departing for Earth in the background. *Photo courtesy McDonnell Douglas.*

This headband-mounted one-inch TV screen is now used aboard space station *Friendship* by our astronauts during repair, assembly, and maintenance work. A voice-activated device, this operations and maintenance information system (OMIS) allows *Friendship* crew members to read instructions and view diagrams while they are working. Information used in OMIS is stored on a laser disk that holds an entire encyclopedia, 54,000 pages, on one side. *Photo courtesy McDonnell Douglas.*

The laboratory in the European Space Agency's Columbus module is based upon their highly successful Spacelab program of the 1980s. It includes 17 materials sciences racks, 15 life sciences racks, and 8 racks of bioprocessing equipment. *Photo courtesy European Space Agency.*

The British Aerospace HOTOL, constructed back in the 1980s, can fly from London, England, to Sydney, Australia, in 45 minutes! This artist's rendering shows the HOTOL in orbital flight high above Earth. *Photo courtesy British Aerospace.*

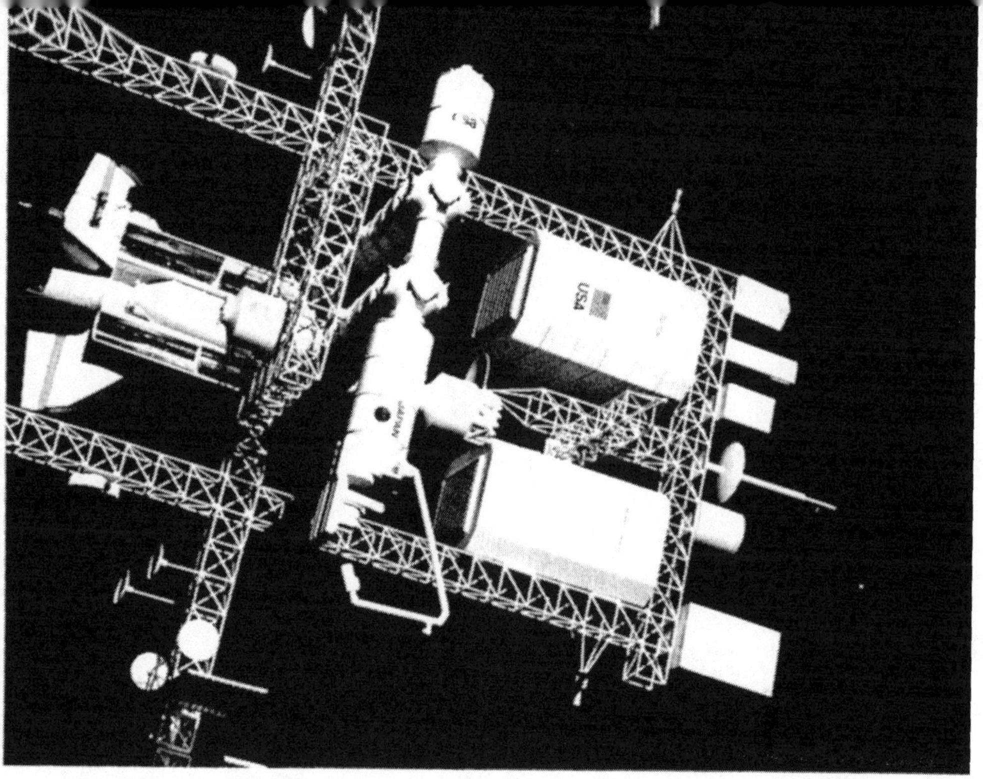

This view of the dual-keel configuration of *Friendship* shows its international flavor. The manned modules in the center include the European Space Agency's Columbus module *(top)* and the Japanese module with its robotic arm for experiments conducted in the vacuum of space. The boxlike structures *(right)* are garages for refueling and repairing satellites. *Photo courtesy McDonnell Douglas.*

An overhead view from a catwalk high above the life-sized full-scale mockup of the space station at the Marshall Spaceflight Center, Huntsville, Alabama. This is one of the locations where the astronauts train before they are flown up to *Friendship* aboard the space shuttles. *Author photo.*

Glossary

Selected Acronyms

ADA   Computer language, named after 19th-century mathematician Ada Lovelace

APU   Auxiliary power unit

ARF   Animal research facility

CERN   French acronym for European Council for Nuclear Research

COAS   Crew member optical alignment sight

CRT   Cathode ray tube

DAP   Digital auto pilot

DT   Deuterium-tritium nuclear fuel

EMU   Extravehicular mobility unit

ESA   European Space Agency

EVA   Extravehicular activity

HMF   Health management facility

ICF   Inertial confinement fusion

IFR   Inertial fusion rocket

JEM   Japanese Experiment Module

JPL   Jet Propulsion Laboratory

kbps   Kilobits per second

KrF*	Gas excimer laser
LM	Logistics module
MEM	Mars excursion module
MIMIC	The Department of Defense's microwave/millimeter-wave-monolithic integrated circuits program
MSC	Manned spacecraft center
MPS	Materials processing in space
NAS	Numerical aerodynamic simulation superconductor system
NSTL	National Space Technology Laboratories
OMV	Orbital maneuvering vehicle
ORU	Orbital replacement unit
OTV	Orbital transfer vehicle
PAD	Project approval document
RMS	Remote manipulator system
TDRSS	Tracking and data relay satellite system
VIP	Fast mode of inertial fusion rocket travel
VOX	Voice-activated communications device

Further Reading

Recommended Periodicals

Aerospace America   Monthly
American Institute of Aeronautics and Astronautics, Inc.
1633 Broadway
New York, NY 10019

Aviation Week & Space Technology   Weekly
McGraw-Hill, Inc.
1221 Avenue of the Americas
New York, NY 10020

NASA Activities   Monthly
National Aeronautics and Space Administration
Superintendent of Documents
U.S. Government Printing Office
Washington, DC 20402

The Planetary Report   Bimonthly
The Planetary Society
65 North Catalina Avenue
Pasadena, CA 91106

Space Education   Every six months
The British Interplanetary Society
27/29 Smith Lambeth Road
London SW8 15Z, England

Space Flight   10 issues per year
The British Interplanetary Society
27/29 Smith Lambeth Road
London SW8 15Z, England

Space World   Monthly
National Space Society
922 Pennsylvania Avenue, SE
Washington, DC 20003

NASA Centers and International Offices

Excellent sources of up-to-date information about the space station are the many NASA centers across the country. The following are those most heavily involved in the space station program, as well as some international offices.

Space Station Office
NASA Headquarters
Washington, DC 20546

Public Affairs Office
Ames Research Center
Moffett Field, CA 94035

Public Affairs Office
Goddard Space Flight Center
Greenbelt, MD 20771

Public Affairs Office
Johnson Space Center
Houston, TX 77058

Public Affairs Office
Kennedy Space Center, FL 32899

Public Information Office
Marshall Space Flight Center
MSFC, AL 35812

CANADA:
National Research Council
Public Affairs Office
Building M-58
Montreal Road
Ottawa, Ontario K1A 0R7
Canada

EUROPE:
European Space Agency
955 L'Enfant North, SW
Room 1404
Washington, DC 20024

JAPAN:
National Space Development Agency of Japan
Suite 570, The Watergate
600 New Hampshire Avenue, NW
Washington, DC 20037

Selected Resources by Chapter

Chapters 1, 2, and 3.
General Space Station Configuration

Cooper, Henry S. F., Jr. *A House in Space,* New York: Holt, Rinehart & Winston, 1976.
Engineering and Configurations of Space Stations and Platforms, NASA Space Station Program Office. Park Ridge, N.J.: Noyes Publications, 1985.
"JPL Research Leads to Possible Cancer Breakthrough." *NASA Activities,* July 1983.
NASA Space Plans and Scenarios to 2000 and Beyond. NASA Office of External Relations. Park Ridge, N.J.: Noyes Publications, 1986.

Perry, Pam. "Cancer Research Advances at I.U. Medical Center." *Indiana Alumni Magazine,* March 1985.

Pogue, William R. *How Do You Go to the Bathroom in Space?* Tom Doherty Associates, 8–10 West 36th Street, New York, N.Y. 10018.

Space Station Program: Description, Applications and Opportunities. NASA Space Station Task Force. Park Ridge, N.J.: Noyes Publications, 1985.

Chapter 4.
Animal House

"Another Dividend from Air and Space: Portable X-Ray Device, the Lexiscope." *NASA Activities,* April 1985.

"Bacterial Enzyme May Be Remnant of Early Evolution." *NASA Activities,* October 1984.

"Bone Stiffness Analyzer Developed by NASA Ames Scientist and Stanford University Engineer." *NASA Activities,* January 1986.

Canby, Thomas Y. "Skylab, Outpost on the Frontier of Science." *National Geographic,* October 1974.

Cooper, Henry S. F., Jr. *A House in Space.* New York: Holt, Rinehart & Winston, 1976.

Covault, Craig. "Spacelab 3 Mission Demonstrates Crystal Growth, Animal Care Challenges." *Aviation Week and Space Technology,* May 6, 1985.

Dixon, Dougal. *After Man: A Zoology of the Future.* New York: St. Martin's Press, 1981.

Gerard, Mireille, and Pamela W. Edwards. *Space Station: Policy, Planning and Utilization.* American Institute of Aeronautics and Astronautics, 1633 Broadway, New York, N.Y. 10019, 1983.

Handley, Douglas O. "Effects of Space on Human System Probed at JPL." *NASA Activities,* September 1985.

"NASA Study Analyzes Need of Humans in Space." *NASA Activities,* March 1985.

Recer, Pau. "Coke Is Delivered Warmly in Space as Pepsi Froths." *Gainesville Sun,* August 15, 1985.

"Skylab Experiments." Vol. 4, *Life Sciences.* NASA and University of Colorado, 1973.

"Space Station Update." *NASA Activities,* May 1983.

Chapter 5.
The Garden

Oberg, James E., and Alcestis R. Oberg. *Pioneering Space: Living on the Next Frontier.* New York: McGraw-Hill, 1986. See especially chapter 8.

"Space Agriculture Researched." *NASA Activities,* May 1986.

Chapter 6.
The Observers

Bossler, John D., and William E. Carter. "Geodesy Goes High Tech." *Aerospace America,* April 1986.

Boyd, L. M. "The Minutiae Man, Rainfall." *Gainesville Sun,* 1986.

DeMeis, Richard. "Building 30 Lives into the Orbital Transfer Vehicle." *Aerospace America,* February 1985.

Emmel, Thomas C. *An Introduction to Ecology and Population Biology.* New York: W. W. Norton, 1973.

"Existence of Polar Wind Proven by NASA Satellite." *NASA Activities,* September 1982.

Friedman, Herbert. "Space Age Lessons About Our Environment." *Aerospace America,* July 1986.

Keister, Edwin, Jr. "Rain or Shine (ECMWF Global Forecast Grid—The European Centre for Medium-Range Weather Forecasts)." *Science Digest,* September 1986.

Mordoff, Keith F. "TRW Designs Orbital Vehicle for Permanent Space Basing." *Aviation Week and Space Technology,* June 2, 1986.

"NASA Studies First Global 'Snapshots' of Ozone." *NASA Activities,* August 1983.

Netter, Thomas W., "Swiss Now Determined to Save Their Nation's Dying Forests." *Gainesville Sun,* March 8, 1984.

Rasool, S. I. "Predicting Earth's Dynamic Changes." *Aerospace America,* January 1986.

Sanders, Deidre, Dick Girling, Derek Davies, and Rick Sanders. "Would You Believe . . . ?" New York: Sterling Publishing, 1974.

Seff, Philip, and David C. Baer II. "Our Fascinating Earth. Dust From Outer Space." *Gainesville Sun,* 1982.

Sitwell, Nigel. "Our Trees Are Dying: Acid Rain." *Science Digest,* September 1984.

"Space Technology: Nicalon Three Dimensional Fibers (OTV Applications)." *NASA Activities,* January 1986.

Sullivan, Walter. "Researchers Draw Back the Curtain on the Mysterious Aurora Borealis." *New York Times,* reprinted in the *Polar Times, Antarctic Journal,* December 1985.

Tyson, Rae. "Hole in Ozone Shield Has Scientists Worried." *U.S.A. Today,* October 21, 1986.

Wilford, John Noble. "Major Finding Supports Theory That Life Began in Clay, Not in the Sea." *New York Times,* reprinted in the *Polar Times, Antarctic Journal,* December 1985.

Chapter 7.
South Pole

Duke, Michael B., Wendell W. Mendell, and Barney B. Roberts. "NASA-JSC toward a Lunar Base." *Aerospace America,* October 1984.

Lenorovitz, Jeffrey M. "Soviet Lunar Polar Mission to Expand Scientific Data Base." *Aviation Week and Space Technology,* March 31, 1986.

"Man to Live on Moon in 50 Years, Panel Says." *Gainesville Sun,* May 20, 1986.

Oberg, Alcestis. "Home on the Moon." *Science Digest,* October 1983.

Oberg, James E. *The New Race for Space.* Harrisburg, Pa.: Stackpole Books, 1984.

Oberg, James E., and Alcestis R. Oberg. *Pioneering Space: Living on the Next Frontier.* New York: McGraw-Hill, 1986.

O'Leary, Brian. "Lunar Base." *Science Digest,* January 1985.

O'Neill, Gerard K. "Building the First Space Colonies." *Future Life,* May 1979.

Pioneering the Space Frontier. The Report of the National Commission on Space. New York: Bantam Books, 1986.

Simpson, Christopher. "Battle for the Moon." *Science Digest,* June 1982.

"Space Committee: Settle on Moon, Then on to Mars." *Gainesville Sun,* May 25, 1986.

Chapter 8.
General Store

Catlin, George. *Letters and Notes on the North American Indians.* Edited by Michael M. Mooney. New York: Clarkson N. Potter, Inc., 1975.

Engineering and Configurations of Space Stations and Platforms. NASA Space Station Program office. Park Ridge, N.J.: Noyes Publications, 1985.

Joels, Kerry Marsh, and Gregory P. Kennedy. *The Space Shuttle Operator's Manual.* New York: Ballantine Books, 1982.

Mails, Thomas E. *Dog Soldiers, Bear Men and Buffalo Women: A Study of the Societies and Cults of the Plains Indians.* Englewood Cliffs, N.J.: Prentice-Hall, 1973.

Reinhold, Robert. "Dining Out—In Space." New York Times News Service. *Cocoa Today,* January 20, 1985.

Chapter 9.
The Industrial Park

Beggs, James M. "Space Commerce—A New Era for Industry." *NASA Activities,* February 1985.

Bird, John. "The First Space Product." *Spaceflight,* November 1984.

Bylinsky, Gene. "What's Sexier and Speedier Than Silicon?" *Fortune,* June 24, 1985.

———. "What Tomorrow Holds." *Fortune,* October 13, 1986.

Clark, Joel P., and Merton C. Flemings. "Advanced Materials and the Economy." *Scientific American,* October 1986.

Covault, Craig. "Unique Products, New Technology Spawns Space Business." *Aviation Week and Space Technology,* June 25, 1984.

———. "McDonnell Douglas, 3M Join to Produce Blood Drug in Space." *Aviation Week and Space Technology,* November 18, 1985.

———. "Spacelab 3 Mission Demonstrates Crystal Growth, Animal Care Challenges." *Aviation Week and Space Technology,* May 6, 1985.

———. "Shuttle Crystal Growth Tests Could Advance Cancer Research." *Aviation Week and Space Technology,* February 25, 1985.

Dumaine, Brian, and Lorraine Carson. "Still A-OK: The Promise of Factories in Space." *Fortune,* March 3, 1986.

Dunbar, Bonnie J., ed. *Materials Processing in Space.* Columbus, Ohio: The American Ceramic Society, 1983.

———. "Space Shuttle: A New Era." In *Materials Processing in Space,* edited by Bonnie J. Dunbar. Columbus, Ohio: The American Ceramic Society, 1983.

"E. W. Systems to Benefit from VHSIC Chips." *Aviation Week and Space Technology,* April 14, 1986.

Engineering and Configurations of Space Stations and Platforms. NASA Space Station Program Office, Park Ridge, N.J.: Noyes Publications, 1985.

"First Space Product Set to Be Developed for Commercial Use." *NASA Activities,* August 1984.

"For Industry It's Almost Lift-Off Time." *Business Week,* June 20, 1983.

Gannes, Stuart. "People at the Frontiers of Science." *Fortune,* October 13, 1986.

Gates, Harry C. "Semiconductor Crystal Growth and Segregation Problems on Earth and in Space." In *Materials Processing in Space,* edited by Bonnie J. Dunbar. Columbus, Ohio: The American Ceramic Society, 1983.

Hendricks, Charles D. "Materials Processing in Space: Inertial Confinement Fusion Target." In *Materials Processing in Space,* edited by Bonnie J. Dunbar. Columbus, Ohio: The American Ceramic Society, 1983.

Horton, Elizabeth. "Can We Conquer Cancer?" *Science Digest,* October 1985.

Jastrow, Robert. "Why We Need a Manned Space Station." *Science Digest,* May 1984.

Jernigan, Camille M. and Elizabeth Pentecost, eds. *Space Industrialization Opportunities.* Park Ridge, N.J.: Noyes Publications, 1985.

"JPL Scientist in Space." *Spaceflight,* October 1985.

Klass, Philip J. "New Microcircuits Challenge Silicon Use." *Aviation Week and Space Technology,* April 16, 1984.

————. "General Electric Centralizes Effort on Gallium Arsenide Technology." *Aviation Week and Space Technology,* January 7, 1985.

Kolcum, Edward H. "Company Plans to Manufacture Crystals in Space." *Aviation Week and Space Technology,* June 25, 1984.

Kreidl, N. J. "Potential Utilization of Glass Experiments in Space." In *Space Industrialization Opportunities,* edited by C. M. Jernigan and E. Pentecost. Park Ridge, N.J.: Noyes Publications, 1985.

"Lewis Research Center Opens Microgravity Materials Science Lab." *NASA Activities,* October 1985.

"Lovelace Studies Cancer Research in Space." *Aviation Week and Space Technology,* June 25, 1984.

Marsh, Alton K. "Honeywell Will Produce Gallium-Arsenide Chips." *Aviation Week and Space Technology,* May 6, 1985.

"Medicine Sales Forecast at $1 Billion." *Aviation Week and Space Technology,* June 25, 1984.

"Microgravity: A New Tool." *Space Education,* May 1986.

"Monolithic Integrated Circuit Effort to Parallel VHSIC Program." *Aviation Week and Space Technology,* April 14, 1986.

NASA Space Plans and Scenarios to 2000 and Beyond. NASA Office of External Relations. Park Ridge, N.J.: Noyes Publications, 1986.

Oberg, James E. *The New Race for Space.* Harrisburg, Pa.: Stackpole Books, 1984.

O'Neill, Gerard K. *The High Frontier: Human Colonies in Space.* New York: Bantam Books, 1978.

Perry, Pam. "Cancer Research Advances at I.U. Medical Center." *Indiana Alumni Magazine,* March 1986.

"Research in Microgravity." *Space Flight,* June 1985.

Rindone, Guy E. "The Universities Space Research Association and Its Role in the Materials Processing in Space Program." In *Materials Processing in Space,* edited by Bonnie J. Dunbar. Columbus, Ohio: The American Ceramic Society, 1983.

Settles, Gary S. "Hidden Frenzy." *Science Digest,* August 1981.
Smith, Bruce A. "McDonnell Douglas Plans to Process Large Pharmaceutical Batch in Space." *Aviation Week and Space Technology,* November 18, 1985.
Solomon, Stephen. "Gallium Arsenide: The Right Stuff." *Science Digest,* November 1982.
Space Station Program: Description Applications and Opportunities. NASA Space Station Task Force. Park Ridge, N.J.: Noyes Publications, 1985.
Sweetnam, George. "The Incredible Shrinking Microcircuit." *Science Digest,* January 1982.
"T.I. Produces 4K Gallium Arsenide Chip." *Aviation Week and Space Technology,* June 10, 1985.
Testardi, Louis R. "Overview of the NASA Materials Processing in Space Program." In *Materials Processing in Space,* edited by Bonnie J. Dunbar. Columbus, Ohio: The American Ceramic Society, 1983.
Uhlmann, Donald R. "Glass Processing in a Microgravity Environment." In *Materials Processing in Space,* edited by Bonnie J. Dunbar. Columbus, Ohio: The American Ceramic Society, 1983.
"Variane Developing Gallium-Arsenide Solar Cell." *Aviation Week and Space Technology,* December 2, 1985.
Wang, T. G. "Review of Containerless Processing Technologies and Facilities." In *Materials Processing in Space,* edited by Bonnie J. Dunbar. Columbus, Ohio: The American Ceramic Society, 1983.

Chapter 10.
StarEagle

Davies, Owen. *The First Starship: The Omni Book of Space.* New York: Kensington Publishing Corporation, 1983.
Dunbar, Bonnie J., ed. *Materials Processing in Space.* Columbus, Ohio: The American Ceramic Society, 1983.
Eyles, Don. "Space Station Thrillers Unfold at Draper Lab." *Aerospace America,* October 1986.
Pitts, John H., Jack Hovingh, and Sam Walters. "Inertial-Confinement Fusion." *Mechanical Engineering,* October 1982.

Chapter 11.
The Curiosity Shop

Bartusiak, Marcia. "Sensing the Ripples in Space-Time." *Science '85,* January/February 1985.
———. "Missing: 97% of the Universe." *Science Digest,* December 1983.
———. "Stars: Ultimate Catalog." *Omni,* February 1986.
———. "The Planet Hunters." *Science Digest,* January 1984.
Baugher, Joseph F. *On Civilized Stars: The Search for Intelligent Life in Outer Space.* Englewood Cliffs, N.J.: Prentice Hall, 1985.
Beggs, James. "Exploring and Working in Space." *NASA Activities,* December 1985.
Bird, John. "NASA's AXAF Space Observatory." *SpaceFlight,* June 1984.

Boslough, John. *Stephen Hawking's Universe.* New York: Quill/William Morrow, 1985.

Boyd, Mary Jo. "Glimpse of Infinity: The Ultimate Telescope Will Take Us to the Edge of the Cosmos." *Science Digest,* July 1983.

Briggs, John. "The Genius Mind." *Science Digest,* December 1984.

Burger, J. J. "A New Eye on the Universe." *SpaceFlight,* February 1985.

Carter, William E., and Douglas S. Robertson. "Studying the Earth by Very-Long-Baseline Interferometry." *Scientific American,* November 1986.

Chaiken, Andy. "A Giant Among Telescopes." *Science Digest,* May 1985.

Chartrand, Mark R., III. "Reading the Script of the Cosmic Drama." *Science Digest,* August 1983.

Dettling, J. Ray. "The Amazing Futurephone." *Science Digest,* April 1982.

"Faraway Planets." *Science Digest,* January 1986.

Fried, Ellen. "The Ungentle Death of a Giant Star." *Science '86,* January/February 1986.

Heidmann, Jean. *Extragalactic Adventure: Our Strange Universe.* Cambridge: Cambridge University Press, 1982.

"Hubble Space Telescope Optics Arrive in California." *NASA Activities,* December 1984.

"Huge, Bright-Blue Arcs Have Astronomers Baffled." *Los Angeles Times,* January 8, 1987.

Huyghe, Patrick. "The Black Hole in Earth's Backyard." *Science Digest,* September 1982.

———. "The Mysterious Fountains of Space." *Science Digest,* March 1982.

"IRAS Astronomical Catalog Available." *NASA Activities,* January 1985.

"IRAS Discovers Its Fifth Comet." *NASA Activities,* November 1983.

Jastrow, Robert, and Malcolm H. Thompson. *Astronomy: Fundamentals and Frontiers.* New York: John Wiley & Sons, 1977.

Lemonick, Michael D. "Gravitational Lens: Mirages from across the Universe." *Science Digest,* April 1985.

———. "Cosmic Clarity." *Science Digest,* May 1986.

———. "Massive Mystery: There's Something Out There Too Big to Explain." *Science Digest,* August 1986.

Lerner, Eric J. "Magnetic Whirlwind: New Discoveries Show That Magnetism Is as Fundamental as Gravity." *Science Digest,* July 1985.

———. "Shaving Angstroms off the Space Telescope's Mirror by Computer." *Aerospace American,* March 1985.

"Mapping the Heat of Heaven." *Discover,* April 1983.

McAleer, Neil. *The Cosmic Mind-Boggling Book.* New York: Warner Books, 1982.

McLaughlin, William. "IRAS and Galaxies." *SpaceFlight,* December 1986.

Mordoff, Keither F. "Company Ends Electrical Tests on Telescope." *Aviation Week & Space Technology,* September 24, 1984.

Nadio, Steve, and Neal Burnham. "Break Dancing in the Pleiades." *Omni,* May 1985.

"New 'Planet' Found." *Science Digest,* March 1985.

Overbye, Dennis. "The Secret Universe of IRAS." *Discover,* January 1984.

———. "The Shadow Universe." *Discover,* May 1985.

Pais, Abraham. "First Word." *Omni,* June 1986.

Pioneering the Space Frontier: The Report of the National Commission on Space. New York: Bantam Books, 1986.

Raymo, Chet. *365 Starry Nights.* Englewood Cliffs, N.J.: Prentice Hall, 1982.

Sagan, Carl. *Cosmos.* New York: Random House, 1980.

"Space Telescope Renamed." *NASA Activities,* November 1983.

Sund, Robert B., Donald K. Adams, and Jay K. Hackett. *Accent on Science.* Columbus, Ohio: Charles E. Merrill, 1980.

Taubes, Gary. "Everything's Now Tied to Strings." *Discover,* November 1986.

———. "Einstein's Dream." *Discover,* December 1983.

Van Flandern, Thomas. "Exploding Planets." *Science Digest,* April 1982.

Waldrop, M. Mitchell. "First Sightings." *Science '85,* June 1985.

"What Is a Quasar?" *Science Digest,* January 1983.

Wolkomir, Richard. "What Are You?" *Omni,* September 1986.

Chapter 12.
Banana Strings and Long Walks

Clearwater, Yvonne A. "A Human Place in Outer Space." *Psychology Today,* July 1985.

———. *A Report on the Activities of the Space Station Habitability Research Group.* Moffett Field, CA: NASA-Ames Research Center, February 1986.

Connors, Mary M., Albert A. Harrison, and Faren R. Akins. *Living Aloft: Human Requirements for Extended Spaceflight* (NASA SP-483). Moffett Field, CA: NASA-Ames Research Center, 1985.

"Effects of Space on Human System Probed at JPL." *NASA Activities,* September 1985.

"Human Rhythm Cycles Studied By NASA Scientists." *NASA Activities,* October 1982.

"NASA Study Analyzes Need of Humans in Space." *NASA Activities,* March 1985.

Taylor, T. C., J. S. Spencer, and C. J. Richa. *Space Station Architectural Elements and Issues Definition Study.* Moffett Field, CA: NASA-Ames Research Center, 1986.

Chapter 13.
Olympus Mons or Bust

Beggs, James M. "The Quest for Mars." *NASA Activities,* August 1985.

Black, Randall. "Taking a New Tour of Mars." *Science Digest,* February 1984.

Carroll, Michael W. "The First Colony on Mars." *Astronomy Magazine,* June 1985.

DiPietro, Vincent, and Gregory Molenaar. *Unusual Martian Surface Features.* Glen Dale, Md.: Mars Research, 1982.

French, J. R., H. N. Norton, and G. A. Klein. "Mars Sample-Return Options." *Aerospace America,* November 1985.

Friedman, Louis D. "Visions of 2010: Human Mission to Mars, the Moon and the Asteroids." *The Planetary Report,* March/April 1985.

Friedman, Louis D., and Alexander Zakharov. "New Robot Missions to Mars." *The Planetary Report,* July/August 1986.

Gore, Rick. "The Planets, between Fire and Ice." *National Geographic,* January 1985.

Haberle, Robert M. "The Climate of Mars." *Scientific American,* May 1986.

Harog, Brian. "Phobos Lander Mars Project." *SpaceFlight,* March 1986.

Joels, Kerry Marsh. *The Mars One Crew Manual.* New York: Ballantine Books, 1985.

Lemonick, Michael D. "Mission to Mars." *Science Digest,* March 1986.

"Mars Observer Investigations Selected." *NASA Activities,* May 1986.

McAleer, Neil. *The Cosmic Mind-Boggling Book.* New York: Warner Books, 1982.

McKay, Christopher P. "The Case for Mars." *The Planetary Report,* March/April 1985.

McLaughlin, William. "Viking Technology." *SpaceFlight,* April 1985.

"A New Look at Alien Landscapes." *Discover,* May 1983.

Oberg, James E. *Mission to Mars: Plans and Concepts for the First Manned Landing.* Harrisburg, Pa.: American Library. Originally published by Stackpole Books, 1982.

————. *The New Race for Space.* Harrisburg, Pa.: Stackpole Books, 1984.

Oberg, Alcestis. "Race to the Red Planet." *Science Digest,* October 1981.

Pioneering the Space Frontier: The Report of the National Commission on Space. New York: Bantam Books, 1986.

Ruoff, Carl, Bryan Wilcox, and Gail Klein. "Designing a Mars Surface Rover." *Aerospace America,* November 1985.

Sagan, Carl. *Cosmos.* New York: Random House, 1980.

————. "The Case for Mars." *Discover,* September 1984.

————. "To Mars: Exploration Justified." *Aviation Week and Space Technology,* December 8, 1986.

"Scientists: Mars Had Great Lakes." *Gainesville Sun,* January 5, 1986.

"Viking: A Martian Era Comes to an End." *NASA Activities,* July 1983.

"Water May Have Played Major Role on Mars." *NASA Activities,* November 1985.

Wilding-White, T. M. *Jane's Pocket Book of Space Exploration.* New York: Collier Books, MacMillan Publishing, 1976.

Chapter 14.
The Shining Princess of the Young Bamboo

Altman, G., G. Ransch, and H. Sox. *Columbus Future Evolution Potential.* Elmsford, N.Y.: Pergamon Press, 1986. Published for the 37th Congress of the International Astronautical Federation, Innsbruck, Austria, October, 1986.

Covault, Craig. "U.S., Europe Deadlock over Station Participation." *Aviation Week and Space Technology,* November 24, 1986.

Davis, Neil W. "Japan's Space Program: A National Priority." *Aerospace America,* March 1985.

————. "Japan Broadens Domestic Role in Satellite Development." *Aerospace America,* February 1985.

Lottmann, Robert V., William E. Rice, and Lynette D. Wigbels. "The International Team: Space Station 1986." *Aerospace America,* September 1986.

Index _____